591.3
R174c

COMPUTERS AND EMBRYOS

Models in Developmental Biology

COMPUTERS AND EMBRYOS

Models in Developmental Biology

ROBERT RANSOM
Department of Biology
The Open University
Milton Keynes, UK

A Wiley–Interscience Publication

JOHN WILEY & SONS
Chichester · New York · Brisbane · Toronto

Copyright © 1981 by John Wiley & Sons Ltd.

All rights reserved

No part of this book may be reproduced by any means, nor transmitted, nor translated into a machine language without the written permission of the publisher.

British Library Cataloguing in Publication Data:

Ransom, R. J.
  Computers and embryos: models in developmental biology.
  1. Embryology
  I. Title
  591.3′307′2   QL955

  ISBN 0 471 09972 4

Typeset by Preface Ltd, Salisbury, Wilts, and printed in Great Britain, by the Pitman Press, Bath, Avon

Contents

Acknowledgements ix

Preface xii

Chapter 1 Introduction 1
  The historical background 1
  Cells and molecules 2
  Embryology as a series of systems 3

Chapter 2 Modelling in biology 7
  Is biology different from the other sciences? 7
  What is a model? 9
  Anatomy of a heuristic model 9
  Advantages and pitfalls of heuristic model building 11
  The variety of heuristic modelling techniques 12
    Paper-and-pencil models 12
    Mathematical models 12
    Computer models 13
    Substitute system models 13
  Descriptive models 14
  Other views of model classification 16
  References 17

Chapter 3 Component processes of development 18
  Developmental parameters 19
    Cell division 20
    Genetic information 20
    Environmental sensing 20
    Differentiation 21
    Movement 21
  Pattern-formation mechanisms 21
  The French Flag 22
  Development in action 24
  Complexity in development 26

The building blocks of development	32
References	33

Chapter 4   Substitute system models 34
The relevance of the substitute system	34
The variety of substitute systems	36
Non-biological substitute systems	37
References	46

Chapter 5   Mathematical models of development . . . 47
Introductory note	47
The spectrum of mathematical models	47
Mathematical models of pattern formation	48
Self-organization in non-equilibrium systems	56
Catastrophe theory and development	61
Conclusions	65
References	66

Chapter 6   Computer modelling and development . . . 68
Computers and models	68
The model simulation	70
Determination and randomness	71
The experimental approach	74
Computers and embryos	76
What can and what can't be simulated?	80
References	81

Chapter 7   Simulating developmental processes: (1) Methods   82
Growth in a one-dimensional row	82
Some basic decisions to be made	84
How many dimensions?	84
Number of neighbours to each cell	85
Two dimensions out of one	86
The edges of the array	87
A two-dimensional cell population growth model	89
Cell division	90
Control lists	90
Changing parameters	94
A model that allows all cells to divide	94
Simplified method	94
The pushing model	94
Bit packing and 'complex' cells	97

Constraining growth 101
A topological model of cell interactions 103
Concluding remarks 104
References 105

Chapter 8 Simulating developmental processes: (2) Models 106
Introduction 106
Subcellular models 107
Automata-theoretic approaches 111
   Early work 111
   Applications 114
   The game of life 115
Cell interaction models applied to specific systems 118
   Cell lineage and interaction in simple organisms 118
   Cancer and models 124
   The vertebrate limb 128
   The aggregation of cells 132
Models of pattern specification 139
Some general notes on computer modelling strategy 152
References 152

Chapter 9 A case study: Computer analysis of insect morphogenesis 157
The development of the fruitfly 159
The insect eye 162
Computer modelling of insect morphogenesis 164
   Running the simulation 164
Results obtained using the model 168
   Average clone patterns produced by S growth pattern . . . 170
   Average clone patterns produced by I and C growth . . . 173
   Individual patterns 174
   Clone growth patterns and relative clone sizes 174
Comparison of practical and computer experiments 174
Updating of the model: simulating leg disc growth 175
   Cells are 'sticky' 178
Postscript 181
References 182

Chapter 10 Developmental modelling and the future 184
Future research 185
   Cells and abstract automata 185

    Computer techniques 185
    Conflict between practical and theoretical studies 186
    Towards a theory of development 187
Concluding remarks 188
References 189

Appendix 1   Video techniques 190
Display of output 190
Advantages of video methods 191
Using a refresh display with a cell population growth program . 191
Use of the light pen 193
Programming the light pen 194

Appendix 2   Further reading 196
Analytical treatments of development 196
'Theoretical embryology' 196
Mathematics and biology 197
Computers 197
Simulation 197
References 198

Appendix 3   Glossary of computer modelling terms . . . 200

Appendix 4   A catastrophe machine 203

Author Index 207

Subject Index 209

Acknowledgements

It would have been impossible to write this book without the help and encouragement of a large number of peers and colleagues both present and past. I owe a particular debt to the late Professor C. H. Waddington, Henrik Kacser, Jim Burns, Dave Scrimshire and Dr N. Xeros. Other help was given by John Oldfield, Peter Allen and John Wexler. Fred Toates kindly read the manuscript and made valuable and encouraging comments. Baz East advised me on figure preparation. None of these people, however, are in any way responsible for errors or misconceptions which may be present. Conversely, where credit is due to other authors I hope I have given sufficient recognition.

I would like to acknowledge the following sources for permission to reproduce material.

The American Association for the Advancement of Science and Michael Wilcox for Figure 3.4, which appeared in *Science*, **191**, 866, 1976 (Copyright 1976 by the American Association for the Advancement of Science); Churchill Livingstone and Tom Elsdale for Figure 3.6; Andrew Tomlinson for Figure 4.1; Dr G. G. Selman for Figure 4.2; A. T. Winfree and Springer Verlag, Heidelberg, for Figure 4.3, from *Biomathematics*, **8**, 313, 1980; Peter Lawrence and Academic Press Inc. for Figure 4.4, which appeared in *Advances in Insect Physiology*, **7**, 203, 1971 (Copyright by Academic Press Inc. (London) Ltd.); H. Honda and Academic Press Inc. for Figure 8.3, from *J. Theor. Biol.*, **42**, 464, 1973 (Copyright by Academic Press Inc. (London) Ltd.); W. Düchting and Elsevier/North Holland Biomedical Press for Figure 8.5, from *Biomathematics and Cell Kinetics* (eds. A. J. Valleron and P. D. M. MacDonald), 1978; The Society for Experimental Biology for Figure 8.14, from *Symposia of the Society for Experimental Biology*, **25**, 379–390, 1971. The Open University also kindly allowed reproduction of Figures 3.7, 8.9, and 9.1 from *S202 Biology – Form and Function* (1980).

I would finally like to thank my wife Anne without whose support and skills this book would never have been either typed or finished.

Preface

This book describes the uses and applications of models in development biology: the study of how adult organisms grow and develop from the single-celled egg. A variety of model systems are discussed in the first five chapters, whilst the second half of the book concentrates on computer modelling. In recent years, modelling techniques have been used in a variety of biological fields to aid the understanding of complex systems, and it is hoped that at least part of the subject matter included here will be of interest to all who are faced with model building and simulation problems in biology. There is often controversy between practical and theoretical biologists about the usefulness of modelling procedures, but as all scientists use models of one form or another I am sure that protagonists of both approaches will find some common ground here.

The chapters may largely be read as essays in their own right, and it is left up to the reader to decide which sections will be of interest in his or her own particular case. The book should be suitable for all levels of reader with a science background and although an elementary knowledge of computer programming is assumed, some help is included for the novice in this area. Those without computer experience are often worried about the mathematical skills involved. I have tried to keep the mathematics involved to a minimum, and even the chapter on mathematical models has been written with the non-mathematician in mind. I hope that the average reader would be able to grasp the essential elements of the various models without any post-High School mathematical training. For this reason, I have also avoided using many algorithms which are in computer language: most explanations are diagrammatic or verbal. In addition, an appendix giving definitions of the computer modelling terms used in the book is included.

I have tried to include the basic developmental biology needed for the reader to understand what the simulations are trying to model, and it should be possible for a non-biologist to follow most of the biological arguments put forward. For the more serious reader, a basic-level university course in biology should be a sufficient prerequisite, especially if the sections on embryology were revised prior to embarking on the present book. The text should be suitable as background reading for tertiary level courses and above on developmental biology, mathematical biology, or computers and biology, and detailed study should take around one-sixth of a semester, together with the associated reading recommended in the text. It would be unfair to suggest that the text has been written with any course in mind, primarily because there has previously existed no text suitable for a course on developmental models, and so no such courses have been devised. Perhaps the present volume may act as a catalyst in getting such a course off the ground.

Many of the methodologies presented are drawn from my own published and unpublished computer simulations of developing systems. This is mainly because so little instructional material regarding computer models and development has so far appeared in print. Interesting simulations and methods can often be found in the abstracts of conferences on computer modelling or cybernetics, but these are usually unpublished and often inaccessible to the scientist outside the narrow field involved. The little information that has so far appeared and can be readily consulted is referenced in Appendix 2, 'Further reading'.

This book covers a variety of model systems, and in describing them it is all too easy to be a jack of all trades and master of none. Although the reader will learn a little about computer programming and even a little about catastrophe theory, I have tried not to get sidetracked from the general topic of models in developmental biology. There are doubtless many models that I have omitted to discuss, and I hope their absence will be forgiven.

R. J. RANSOM

Chapter 1

Introduction

THE HISTORICAL BACKGROUND

A basic tenet of developmental biology—the general name given to the process by which a single-celled egg develops into an adult organism—is that of simplicity giving rise to complexity. The human egg, for example, produces around 10^9 cells by the time adulthood is reached: these cells are of many different types arranged in patterns to allow a variety of activities like movement (muscle cells), thought (brain cells) or protection (skeletal cells). How does simplicity result in complexity during development? This question has plagued embryologists since Aristotle, because it is difficult to see how something so seemingly 'simple' as the

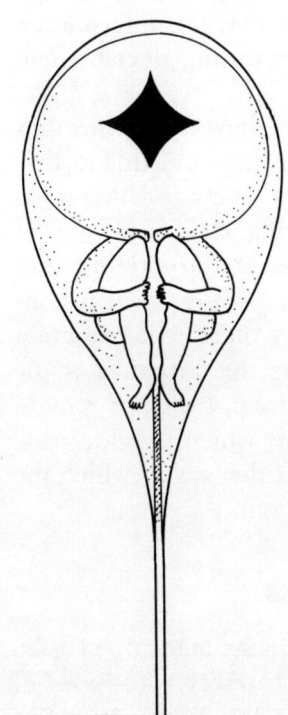

Figure 1.1 Drawing of a human spermatozoon. (After Hartsoeker.)

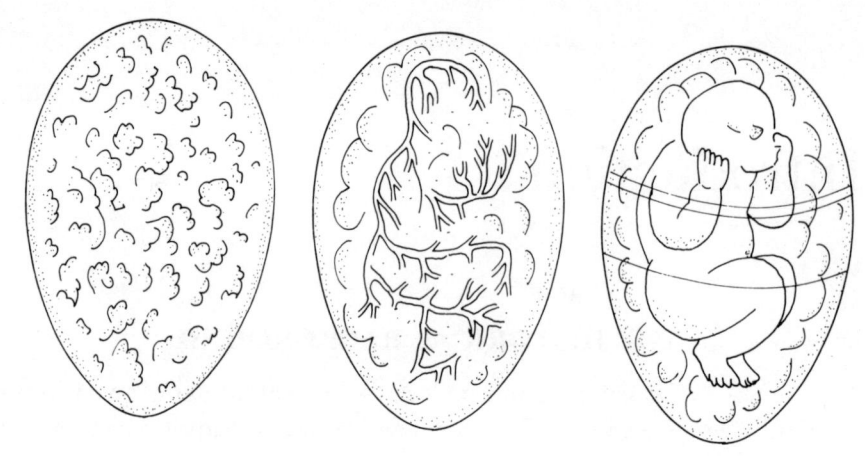

Figure 1.2 Aristotelian view of development. (After Jacob Rueff, 1554.)

egg can produce a complex, mature organism. The dilemma was inadequately explained by the preformationists, who held that a complete human foetus was curled up in the head of each human sperm cell (Figure 1.1). Conversely, Aristotle had previously argued that the egg held only limited developmental information, and that this information itself generated new properties and complexities during development (Figure 1.2).

Microscopical and biochemical observation has shown that the adult certainly does contain more information than the egg, and this at first sight is evidence for Aristotle's theory. But how does the increased amount of information in the adult come into being?

The truth is probably somewhere between the preformationist and Aristotelian extremes: the *blueprint* to make an adult organism *is* contained in the egg, but it is arranged in such a way that the information can only be unlocked in the correct sequence as the structure of the growing embryo progresses through time. In addition, the developmental blueprint must interact with the environment, which provides raw materials and stimuli, and these in turn will affect the way in which the component processes of development are carried out.

CELLS AND MOLECULES

With the nineteenth-century findings that organism are made up of cells, and that the chromosomes are the material of heredity, the bases of modern developmental biology were laid. In the last forty years, our knowledge of the biology of molecules has unravelled many fundamental biological problems: the nature of the genetic code, the mechanism of muscle contraction, the structure of many cellular components. Are we close to being able to describe development in equally precise terms? What kinds of processes must we analyse in order to understand development?

Cells consist of a nucleus and cytoplasm. The nucleus contains the chromosomes: entities made of nucleic acids which normally duplicate on every cell division so that each cell keeps a copy of the complete chromosome set. The chromosomes interact with the cytoplasm to produce developmental change, and this process is repeated until the adult organism is formed. There are a series of intra- and intercellular levels at which control can take place. Because we know the genetic code and its relationship to the manufacture of chemical substances in the cytoplasm, we can say something about how particular substances are formed in particular cells (for example, salivary enzymes in salivary gland cells). In some cases we can even describe how the production of particular proteins is controlled at the chromosomal level.

There are only isolated examples which it is possible to analyse in this way, and they tend to be restricted to simple organisms like bacteria. The fact is that even the three dimensional structure of chromosomes in higher organisms eludes detailed description: this suggests that the analysis of a complex process like formation of a limb or brain may be a long way off.

On the other hand, do we need to know every last detail of a process to understand how it works in broad terms? Organisms are highly complex, and it does not require an extensive biological knowledge to appreciate than an overall description of a process like brain development in molecular terms would be extremely complicated. 'Description' can mean different things depending on the level at which a particular process is being studied. Can we rationalize the various levels of activity of a biological process?

EMBRYOLOGY AS A SERIES OF SYSTEMS

There is obviously a need when analysing any complex process to be able to break the process down into the simplest components that will allow a picture of the process to be built up. In the case of any mechanical process, this is a relatively simple matter, because the parts will be recognizable components such as gear wheels, axles, metal plates, the majority of which will usually be needed for the correct operation of the process. Perhaps a few screws or nuts can be omitted, but the machine will not function in the absence of any major components.

Biological systems are more difficult to analyse in this way. If we look at a particular organ or tissue, we can immediately see that the general structure is a function of the layout of the cells within the tissue—remove some of the cells and the tissue will collapse. What of the make-up of the individual cells? Is every component of every cell important in every activity of the tissue? If not, how can we separate out the various activities which go on? The cellular reactions controlling homeostasis in a cell are different from the reactions important in morphogenesis or differentiation, and some method of logically separating these activities is required.

One possible approach is to use the concept of *systems*. A system is an assembly of parts or components connected together in an organized way. A cell could be considered as a system in *molecular* terms, by looking at the working of the genes and the biochemical activity inside the cell. A cell could also be considered as a structural system made up of organelles and with a morphology that is capable of changing on differentiation. At a higher level, in the *cellular* system, cell to cell interaction leads to the formation of *organs* (for example, a limb consists of bone, cartilage, skin and muscle cells). Finally, the whole collection of organs comprises the adult organism, a system in its own right.

The job of description is simplified by the systems approach. Taking a particular example, we could examine the important developmental questions which are relevant at each system level. Figure 1.3 shows an

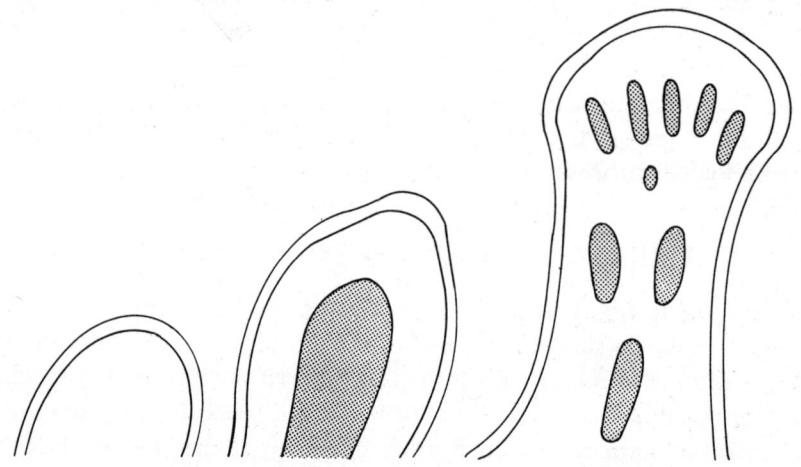

Figure 1.3 Condensation of cartilage elements of a vertebrate limb bud.

early stage in the growth of a vertebrate limb bud, which will go on to form the adult limb. One of the central problems is how the cartilage cells form the skeletal elements of the limb. Let us consider what happens at the organ, cell, and molecular level.

Organ level: The important factors here are those determining pattern and shape of the limb. What mechanism is responsible for ensuring that cartilage cells differentiate in the right place? Why is the overall limb shape always produced?

Cellular level: How do cells interact with their immediate neighbours? How does differentiation from precursor cell to cartilage cell take place—what changes do the cells go through during differentiation?

Molecular level: Which genes are active in the cartilage cells and how are they activated? What biochemical products do the differentiated cells make?

In this book we will be looking at development as a series of systems, concentrating largely on the cellular and organ levels. The next chapter considers how biological systems can be modelled, and looks at the variety of model types that are in use. Chapter 3 takes a more detailed look at the problems of developmental biology and tries to put them into systems language. The next three chapters take particular types of model, giving examples and appraisals of their relative merits. Chapter 4 considers substitute system models, where simple developing organisms model general growth processes. Chapter 5 investigates mathematical models, whilst Chapter 6 gives a general introduction to computer modelling.

Chapter 7 describes strategies for computer modelling of developing cell populations. The accent here is placed on array-bound computer models that allow the spatial components of development to be simulated. Chapter 8 examines a variety of computer models in developmental biology that have been published in the last twenty years. Such models have appeared more sparsely in the literature over the past five years, and the economic climate of the mid- and late 1970s, with the concurrent research emphasis switching to practical projects with more foreseeable short-term results may well have been an important factor.

Chapter 9 describes my own simulations of the development of the fruitfly *Drosophila*. It is hoped that by presenting a case history at length it will be possible for the reader to obtain more of a 'feel' for the abstract way in which other parts of the book are set out. The final chapter looks at the future, and suggests how advances in computer methods, and a more radical approach by researchers, may lead to a more beneficial and efficient usage of model building in developmental biology. Four appendices are also included. The first deals with video methods for building computer models and observing their growth on video display screens. This is of interest because of the dynamic nature of computer models of growth. The second appendix lists books which may be of interest to the reader who wants to find out more about the various types of models and development. The third appendix is a glossary of modelling terms used in various places throughout this book: it is hoped that this will be useful for the reader with little or no knowledge of models, computers and computer jargon.

A final appendix looks at the construction of a simple 'catastrophe machine' which may help the reader to understand the way in which catastrophe theory is used to describe complex phenomena.

A large proportion of this book is dedicated to computer modelling, and I make no apology for this fact. I believe that the complexities of development may be best studied by computer analysis, although some systems are more amenable to this sort of approach than others. The reader will be able to make up his or her own mind about the validity of this argument and I will feel gratified if even a modest interest in developmental simulations is aroused by the material included in the following pages.

Chapter 2

Modelling in biology

IS BIOLOGY DIFFERENT FROM THE OTHER SCIENCES?

Biological processes are normally highly complex in character. Whereas we can often treat systems in physics and chemistry in a direct mathematical matter, biology does not easily lend itself to such an analysis, a state of affairs fueled by the fact that in the past most applied mathematics has been devised to provide a conceptual basis for understanding the workings of *non-living* systems. Certain branches of mathematics, for example differential equations, recurrence relations, and probability theory, are of use in the analysis of certain aspects of biological systems, but it is only rarely that a full description of a dynamic living process can be given in the language of classical mathematics (see Chapter 5). The more complex a system, the more important it is to have a framework in which to give some kind of coherence to the information we have on the system—without such a framework we will not be able to handle the logical complexity needed to understand how the system works (Ransom, 1973).

It is, therefore, important to know why there are additional problems in treating living organisms in the same way as inanimate processes. An insistence that the study of organisms requires the consideration of additional factors to those of the physical sciences does not imply a dualistic or vitalist view of nature; living beings have been affected for some three billion years by historical processes which have had innumerable effects on them, most conspicuously on their genetic programmes. An additional concept is that of organization. Living systems seem to organize themselves into individual units, generating patterns in time and space which are recognizably individual and complete; the basic unit of these patterns is the cell, and the cellular multiplicity of most organisms gives additional complexity. It is this organization, the stability which accompanies it, and the recording and passing on of the organizational history by means of self-replication which results in the overall complexity of biological systems. The central argument of this book is therefore that modelling is necessary to achieve an understanding of such intricate processes.

Perhaps the biggest difference between living and non-living systems is the genetic programme possessed by living cells, although the ability of biological structures to refine their activities by means of evolutionary strategies does not concern us here and has been reviewed extensively elsewhere. Besides their role in heredity, the genes also play a central part in the control of development, and any developmental model must take stock of the importance of the genes.

Any attempt to model development must also recognize the peculiar hierarchical nature of biological systems. Genes control the synthesis of biochemicals within individual cells, but it is cellular interactions that determine biological shapes and patterns. The relationship between genes and cell interactions remains unclear because of the difference in levels at which they occur. Gene action is explainable in terms of biochemistry. Cell–cell interactions are not analysable in the same terms because of the millions of interactions that occur within and between cells: how do we isolate a single 'cell interaction' between two neighbouring cells?

To some extent the systems approach outlined in the previous chapter may help. By dealing with a single systems level, the incompatibility of different levels on the biological hierarchy can be avoided. There are other pitfalls to trap the unwary, however. Finding out more and more about how the component parts of a system are made up may tell us little about how the system as a whole works. An example often quoted is that of the watch: given a mass of springs and cogwheels, it would be difficult for someone unfamiliar with the workings of the watch to explain how it functions. A metallurgical examination of the structure of the cogwheels is clearly not of great value—it would be far better to try to piece the various parts of the watch together, and see how they work in conjunction with one another. The metallurgical study may be very interesting, but it is of rather secondary importance. This difference between parts and whole has been discussed by several previous authors. Weiss (1963) has used the argument

> 'If a is indispensable for both b and c; b for both a and c, and c for both a and b, no pair of them could exist without the third member of the group ... in other words, a system of this kind can exist only as an entity or not at all'.

At the lowest levels, the properties of the whole system may not be evident, and it is only by looking at the whole that the working can be seen.

Apter (1966) quotes the so-called 'levels of organization' doctrine.

'The living organism can only meaningfully be dealt with at a higher level than physics and chemistry, and at this level quite new phenomena emerge which are of the very essence of such self organizing systems and are missed at any lower level.'

Kacser and Burns (1967) have also discussed an approach of this kind, and warn of the dangers of the fine analysis of trivial phenomena,

'a new methodological and conceptual approach is necessary if we are to make progress beyond the cataloguing of an even greater number of biochemical details'.

WHAT IS A MODEL?

If we have knowledge about the individual parts of a biological system, but no detailed information about the interrelation of the parts, then it may well help to build a model of the system. What *is* a model? The word 'model' has been subjected to a variety of alternative usages in biological research. In its simplest form, a model is a simplified representation of a structure. In turn, the word 'structure' may confuse: in this context it stands for an object or system. The Eiffel Tower is an object, and the brain, involving both classes of objects and their dynamic interrelations is a system. A *heuristic* model is a model used to discover how a process works rather than being a *descriptive* model of the process.

A model of the Eiffel Tower therefore involves a static representation, as would a child's plastic 'model' of an airplane or ship. It is basically a descriptive model. A model of the brain is more complex, and would probably include provision for modelling the dynamics of brain interactions. It would probably be heuristic in character. The definition of a heuristic model is in fact rather simple, but it is the way in which such models are constructed that gives rise to most difficulty in classification. Static models can be heuristic; for example, the form of a protein molecule is often worked out by trial-and-error construction of plastic models or an intricate matchstick model correct in every detail.

ANATOMY OF A HEURISTIC MODEL

To a certain extent, model building is an amateur science amongst professional pure scientists. There is no doubt of its importance, but

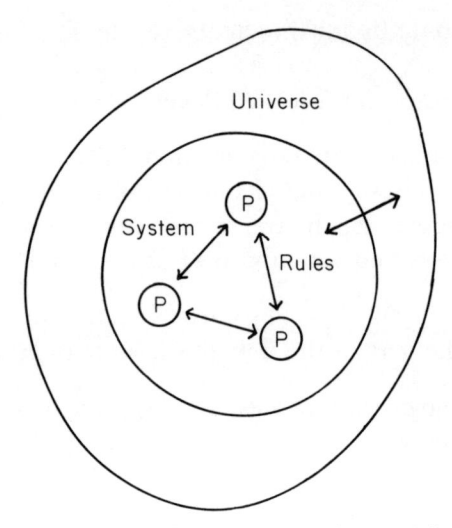

Figure 2.1 The elements of a model.

most scientists have no formal training in modelling and have no desire to be educated in what they often see as an abstract exercise with no direct relevance to their research. This state of affairs may be constrasted with the situation in technological and applied science where modelling strategy is commonly taught to students. The more nebulous nature of models in pure science, and especially biology, makes it even more imperative that models should be analysed in detail. A dynamic model normally consists of the elements shown in Figure 2.1.

The *universe* consists of the entity to be modelled, be it growth of a limb, the Eiffel Tower, or the theory of evolution, together with its relationship to other entities. The universe of a limb therefore includes the rest of the organism, as the organism provides nutrients for cell growth. The Eiffel Tower must rest on the soil, and the air and soil are the external components or the universe in this case. The theory of evolution is more complex, its universe being, perhaps literally, our own Universe itself, as the formation of the Earth and creation of life are essential elements of evolution and natural selection.

The *system* is made up of the particular entity to be modelled in isolation. Limb growth and the Eiffel Tower are obvious systems. The theory of evolution is represented by living organisms—their morphology and distribution. We saw in the previous chapter that development may be broken down into a series of different systems levels.

The *parameters* are the component parts of the system: its nuts and bolts. One of the major problems of modelling is to sort out the particular nuts and bolts to be modelled. The most important parameter of the theory of evolution is natural selection. The Eiffel Tower is made of steel girders. In the case of the limb, some of the parameters are obvious, like cell size and shape, whilst the nature of the control signals for specifying shape and growth of the limb are more problematical. If we knew the parameters of development, our modelling task would be infinitely easier.

The *rules* are simply the interactions between the component parts of the system. If we identify natural selection as an important parameter, then the interaction of fit animal X and unfit animal Y must take place according to set instructions. The 'rules' of the Eiffel Tower are the construction methods. The limb has far more complex construction methods, and it is their elucidation that represents the problems and challenges of developmental biology.

ADVANTAGES AND PITFALLS OF HEURISTIC MODEL BUILDING

We have not yet considered, except in a rather roundabout way, *why* models should be constructed. The major reason is one of complexity. We have suggested that if a biological system is broken down into a number of individual parts it is difficult (and often impossible) to hold a clear mental picture of how the various parts interact to produce a particular working organ or tissue. At a physiological level it is often possible to describe the function of the adult tissue. We know the micro-anatomical structure and physiological processes that compose the adult vertebrate kidney, for example. A more complex problem would be to consider how the kidney develops. This is a higher-order question, because it involves not only physiology and anatomy but also the developmental interrelations between the two. (The various component processes of development will be the subject of the next chapter.)

Use of modelling becomes more and more necessary the more complex is the real system under examination, and the extra problem of changes in structure with time rank developmental biology along with the brain sciences among the most difficult of biological problems to unravel. The two may well be related problems. If we could work out the rules for brain development, we might be much closer to an explanation of how neutral circuits function. This is a side issue for the present, but leads us yet again to the argument that a system is more than the sum of the component parts. A dynamic model—either a computer or 'substitute system' model—will allow investigation of the relationship between the parts, and the reliance of the whole system on its individual components. There are both good and bad models, however, and it will often be found that use of an irrelevant model will lead the researcher into an endless sea of red herrings. This 'relevance dilemma' will be further investigated in Chapter 4. A good model will have as many of the features of the original system as are necessary to identify how the original system works. This seems at first reading to be a tautological statement, and has been possibly summed up more accurately by Goodwin (1970):

> 'The choice of (model) language, of formalism, lies before the investigator, and the only rule for this choice is the general one that the good butcher does the least cutting.'

THE VARIETY OF HEURISTIC MODELLING TECHNIQUES

Paper-and-pencil models (static models)

Models of this type are the most straightforward, and have been used by all types of scientist throughout history. Drawing successive stages of limb growth, for example, may give clues as to how shape and form change during development. The classic example of the application of this type of modelling procedure to developmental biology can be seen in D'Arcy Thompson's book, *On Growth and Form*, published over seventy years ago. By comparing biological shapes, Thompson was able to lay the foundations for modern studies of morphogenesis. The drawback of models of this type is their lack of provision for treating dynamic processes, a difficulty surmounted to some extent by using them in conjunction with mathematical models, which are considered in Chapter 5.

The other categories of model could all be largely considered as dynamic models as they involve some aspect of the temporal interactions within developing systems. Some mathematical models are descriptive, however. An example is the work of D'Arcy Thompson himself, discussed in Chapter 5.

Mathematical models

Mathematics itself can be regarded as an example of a model language in certain instances, although we have already seen earlier in this chapter that its direct use in biological situations may be limited. In the cases in which it can be used, classical mathematical equations are constructed (or more often, adapted) and are used to describe a process. An example of a 'mathematical adaptation' in biology is the use of statistical mechanics, first devised to predict the thermodynamic behaviour of gases, and recently used by Goodwin (1962) to formalize subcellular dynamic activities. In this type of model, adjustments in input parameters to the model can often be used as a test to see if the model is behaving like the real system. The *discrete*, statistical or probabilistic model is another type of mathematical model. However, in this case, the elements of the system being studied are too complex to fit into the regular description given by a continuous mathematical model and models of this type usually manifest themselves in the form of computer models. Continuous mathematical models are discussed in Chapter 5.

Computer models

Computer models can often take elements from both paper-and-pencil models and mathematical models to produce hybrid models, normally 'animated' as *simulations*. A better definition of a simulation might be 'the dynamic representation of a model on a computer'. How can a paper-and-pencil model be simulated? Imagine that the bottom right-hand corners of the pages in this book have sequential drawings of the growth of a vertebrate limb on them. If the pages are flicked through quickly, a representation (or simulation) of the growth of the limb will be observed. If the drawings were out of register, a 'garbled' growth process would be seen. The sequential representation of a process at different states in time is the essential basis of the computer model, except that the computer performs rather more sophisticated versions of the 'page-flicking' operation which may be more or less similar to the real system. Various types of computer models exist, and these will be dealt with in detail in Chapters 6 to 9.

Substitute system models

In many cases, a biological system is in itself used as a model, whilst in other situations a non-living physical analogue may be employed. Many aspects of development in both animals and plants are studied by using organisms much simpler than those in which our interest might be centred—the interaction of two different biological tissue types may be studied *in vitro* using a narrow-meshed filter, or someone interested in cell–cell signalling in higher organisms may study the chemical impulses leading to aggregation in simple slime moulds. Electrical networks can be used as analogues to model nerve nets, and a membrane physiologist may use collodion films and silica gels as analogues of real membranes. Although these have the same functional properties, there is no physiochemical equivalence. Models of this type will be discussed in Chapter 4.

> 'In a sense, it is correct to say that the study of biological systems in terms of models proceeds by abstracting the physics and chemistry from the system, leaving behind a set of formal relations that may then be studied' (Rosen, 1968).

DESCRIPTIVE MODELS

The major part of the present chapter has been concerned with the study of heuristic models. Descriptive models will play a relatively minor role in this book, although it is important to be clear about their nature. Descriptive models in biology tend to be 'paper-and-pencil' models, and although this title sounds a little crude, some intellectual *tours de force* fit into this category. Darwin's theory of evolution is a clear example. By taking a large collection of data regarding the morphology and distribution of animal groups, Darwin was able to synthesize his model of how natural selection controls evolution. One of the most prolific developmental modellers whose work also involved 'paper and pencil' models was C. H. Waddington. Waddington was originally trained as a palaeontologist and had a rather non-mathematical brain ('after all, I am a biologist: it is plants and animals that I'm interested in, not clever exercises in algebra or even chemistry'). Nevertheless, Waddington was able to extract the important elements of many developmental systems and put them into terms which humbler biologists were then able to analyse. Much of his talent is displayed in *Principles of Embryology* (1953), now sadly out of print. A more recent publication is his *New Patterns in Genetics and Development* (1962), esential reading for any model-orientated biologist.

A good example of a Waddingtonian model may be found in his 'gene-action system' (Figure 2.2). This is an attempt to show the relationship between genes and phenotype (the observable characters) of an organism. This model suggests the complex interactions that must go on during development both between various levels (for example, gene–protein or protein–cytoplasm) and between different constituents at the various levels (for example, gene–gene). Waddington envisaged groups of gene-action systems concerned with the differentiation of a

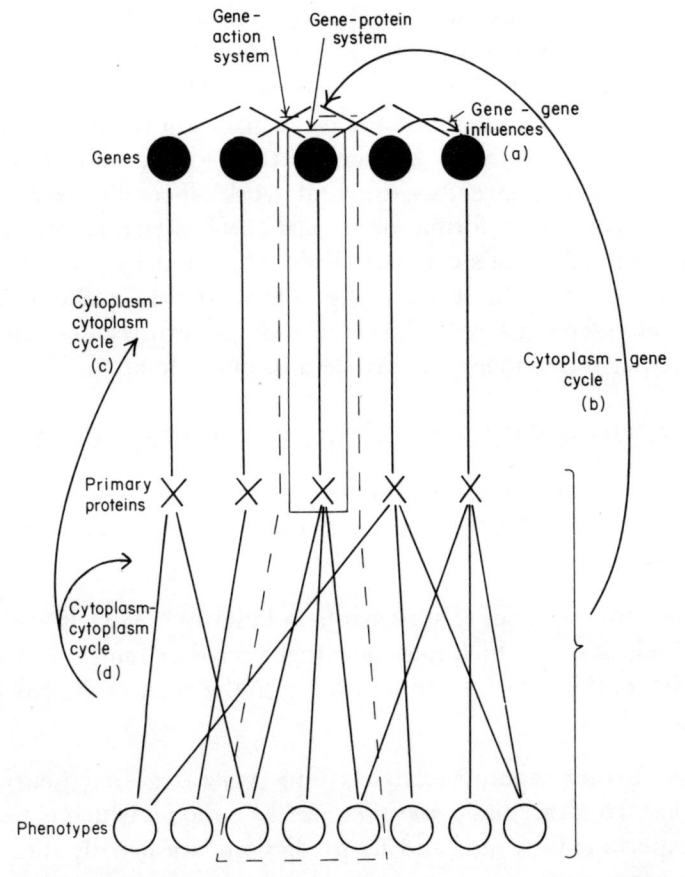

Figure 2.2 Waddington's 'gene-action system'. Each gene determines a primary protein, and the proteins interact with one another to produce the final phenotype, the observable characteristics of the organism. The group of processes connecting gene and final phenotype is called a gene-action system (shown here in dotted lines). (Waddington, 1962.)

particular cell type to be interlocked with one another so as to form a series of buffered or stabilized developmental pathways in the organism. He called these pathways 'creodes'.

A model of this type is abstract, but of immediate applicability to developmental studies. By looking at how particular systems in particular developing organisms might fit into the scheme, the researcher is immediately channelled into devising experiments to test the model's predictions. Waddington was not the only important 'paper-and-pencil' modeller in the history of developmental biology. Many of the great embryologists of this century have tried their hand, albeit less prolifically, at modelling, and another example is Child's 'double gradient theory' (1941) which predicted the causal basis of pattern formation in embryos (unproven as yet). A second pattern formation model that has done much to influence experimental work since the early 1970s is Wolpert's 'positional information' model (1969), itself largely a restatement of many of Child's original ideas. This model is most frequently couched in terms of the 'French flag'—how would a cellular organism form the characteristic red white and blue pattern seen on the French flag? Chapter 3 considers this problem in more detail.

OTHER VIEWS OF MODEL CLASSIFICATION

The above 'working classification' of models is not the only one that exists, and indeed, definitions and categorizations abound. Kacser (1960) describes a model as

> 'a statement, or series of statements in language. Models are therefore propositions, which may be either verbal or mathematical ... in which entities are related according to the rules of the particular language.'

Kacser further separates models into two categories, heuristic and conceptual. He describes a *heuristic model* as a model which is used as an aid to experimental work, and its predictions are usually directly testable by experiment.

> 'It need not be a mathematical model, and it need not make quantitative statements, nevertheless, it contains statements about the outcome of future experiments or observations, by means of which it can be tested for its applicability.'

Most biologists build heuristic models; indeed, experiments are usually designed as a result of such modelling. A model may be simply a means of classifying a group of related plants, a way of describing the interrelation of the components of a biosynthetic pathway, or a tentative verbal explanation of the workings of a physiological process.

A *conceptual model*, on the other hand, acts as scaffolding into which existing knowledge can be slotted, and an overall understanding of a particular system may be thus obtained. Conceptual models often correspond to 'theories', for example, the theory of evolution. Kacser emphasizes that some overlapping may take place between model types.

Other authors have also classified models similarly, including Apter (1966) and Apostel (1961), although differing nomenclature may be used.

Developmental biology suffers from a surfeit of heuristic models. The central problem is to construct suitable conceptual models that will enable us to describe embryology in helpful terms. We have very little idea of how to go about this, except in a very piecemeal fashion. Some scientists believe that theories result directly from experiments, whilst others maintain that an intermediate stage must involve detailed heuristic model construction. The reader must decide which of these viewpoints he will take. Much of the popularity of the 'direct' approach, however, is due to ignorance of the value of modelling techniques, although workers on both sides realize the complexity of developing systems. It is hoped that this book will act as a guide to the neophyte modeller and will help him to put the various modelling techniques into perspective.

REFERENCES

Particular model types will be referenced in the relevant chapters to follow, but interesting discussion of the general use of models in biology may be found in the papers by Kacser and by Goodwin, and in the book by Apter.

Apter, M. J. (1966) *Cybernetics and Development*, Pergamon Press, Oxford.
Apostel, L. (1961) Towards the formal study of models in the non-formal sciences, in: *The Concept and the Role of the Model in Mathematics and the Natural Social Sciences* (H. Freudentel, Ed.), Reidel, Dordrecht, Holland.
Child, C. M. (1941) *Patterns and Problems of Development*, University of Chicago Press, Chicago.
Goodwin, B. C. (1962) *The Temporal Organization of Cells*, Academic Press, N.Y.
Goodwin, B. C. (1970) Biological stability, in: *Towards a Theoretical Biology 3. Drafts* (C. H. Waddington, Ed.), Edinburgh University Press.
Kacser, H. (1960) Kinetic models of developmental and heredity, in: *Symposia of the Society for Experimental Biology*, **14**, Models and Analogues in Biology, 13–27.
Kacser, H. and Burns, J. A. (1967) Causality, complexity and computers, in: 3rd International Symposium, Biologische Anstalt Helgoland, *Quantitative Biology of Metabolism* (A. Locker, Ed.).
Ransom, R. (1973) Can a computer grow limbs? *New Scientist*, **59**, 385–387.
Rosen, R. (1968) Recent developments in the theory of control and regulation of cellular processes, in: *International Review of Cytology*, **23**, 25–88.
Waddington, C. H. (1962) *New Patterns in Genetics and Development*, Columbia University Press, N.Y.
Weiss, P. (1963) The cell as unit, *Journal of Theoretical Biology*, **5**, 389–397.
Wolpert, L. (1969) Positional information and pattern, *Journal of Theoretical Biology*, **25**, 1–49.

Chapter 3

Component processes of development

Now that the philosophy behind modelling has been outlined, we turn to the subject matter the techniques can be applied to. Problems in development are largely problems of cell interaction. In a mature organism, cells sit side by side and cooperate in a regular, definable manner. For instance, cells of the dermis secrete the body's skin covering, and epidermal cells of the small intestine may produce enzymes to enhance food digestion. Because adult cells do not undergo the complex changes of development, their interactions are correspondingly simpler—the shapes and patterns of the organism have already been formed, and most cells only divide to replace those lost by normal turnover. The only problem of cell interaction that really concerns us at that stage is cancerous growth, the subject of several models discussed in Chapter 8.

In development, the accent is on the dynamic interrelations of neighbouring cells. Generally speaking, multicellular organisms develop from a single cell called the egg. A number of growth instructions are held in genes in the chromosomes of the egg and in its cytoplasm. After receiving a fertilization signal and the other half of the chromosomes of the future organism from a sperm, the egg starts to divide into daughter cells. The cells are not terminal, differentiated structures but primordial units which must change and develop according to their environments and their chromosomal, inherited lineage. How, then, is an adult organism formed?

The absolute answer to this question is still unclear, and of course forms the basic material of this book, but we know enough now to phrase it in more specific terms. What, and how, are the 'messages' received by cells from their environments during the development of the embryo? How are these messages interpreted by the chromosomes of the cells? How are different cell types generated? How are biological shapes and patterns formed?

DEVELOPMENTAL PARAMETERS

The main unit of multicellular development is the cell. Each cell can exist in a number of states, depending on time and place in the growing organism. The states are changed by input from neighbouring cells and by reference to the instructions contained in the genetic material in the chromosomes. If a cell exhibits particular characteristics, it is said to be *differentiated*. If a cell is undifferentiated but preprogrammed to give rise to a differentiated structure, it is *determined* to form that structure.

The relationship between differentiation and determination and the reversibility of the two states under experimental stimuli, pervades the literature of developmental biology. It has been shown, for example, that certain insect pre-adult cells can *trans*determine into structures normally formed by other cells, if the pre-adult cells are allowed to undergo extra cell divisions by experimental manipulation. We also know that determination and differentiation are stepwise phenomena progressively building up the complexity needed to form the adult organism. The way in which particular cell groups are separated to form certain organs is not yet clear, although analysis of the fruitfly *Drosophila* has shown that whole groups of primordial cells may be genetically switched to form, for example, wing cells during early development, and these cells are capable of maintaining fixed borders which cannot be invaded by cells of a different genetic type.

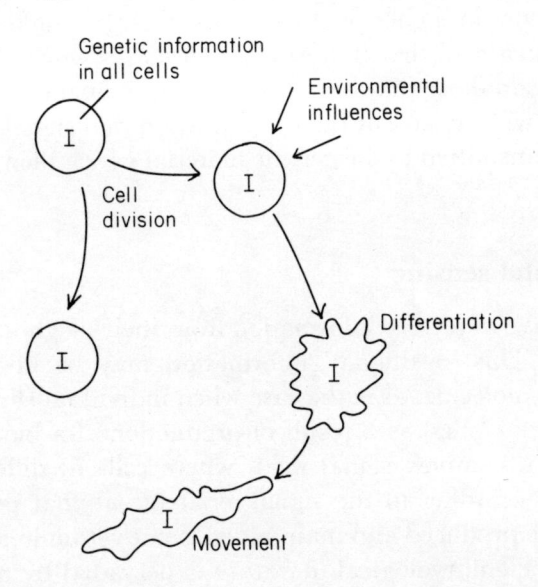

Figure 3.1 Important properties of cells during development.

A list of the important properties possessed by cells during development would include the following, and is shown in diagrammatic form in Figure 3.1:

Cell division
Information held in genetic code
Ability to sense information for environment (i.e., neighbouring cells)
Ability to change state ('differentiate')
Movement.

Cell division

This is important for two main reasons. Firstly, because it provides the motive force for differentiation which normally occurs only after cell division. Also, an increase in cell number changes the morphology of the whole organism and so may affect the input received by any cell.

Genetic information

Initially the genetic information is present in the form of nucleic acid molecules in both egg cytoplasm (the non-nuclear cellular matrix) and nucleus. The cytoplasmic nucleic acid information is rapidly used up in early development, leaving the chromosomal nucleic acid material as the central control element of the embryo. How this control works at the level of the genes in higher organisms is still largely unknown, although it has been suggested that batteries of genes are switched on and off by similar means to those that have been shown to operate in bacteria. For our purposes we may accept the simplification that signals impinging on the cell are transmitted to the genetic material which then acts according to the signal.

Environmental sensing

Cells often have to receive information from their neighbours in order to differentiate. This 'positional' information may be in the form of a simple signal molecule, as is the case when individual slime-mould cells aggregate into a 'slug' as a result of production of a biochemical, or it may involve a complex signal field, where cells at different positions differentiate according to the signal available at that point. Just how such fields are produced and maintained is not yet understood in practical terms. The embryological literature is pervaded by notions of 'gradients of developmental capacity' and some sophisticated theories have been proposed although they avoid detailing the physical bases of the signalling mechanisms. Some of these ideas will be expanded on in Chapters 5 and 8.

Differentiation

We have already noted that any 'state change' within cells may be termed differentiation (although a limitation to *observable* changes is normally used), and the term 'determination' defines the changes within the cell (mainly at the chromosomal level) which will *eventually* result in observable differentiation. Differentiation may involve cell shape changes, production of signals transmitted to other cells, or production of a new biochemical not previously synthesized by the cell.

Movement

Cell movement occurs passively as cells grow and divide, but many types of cell have genuine motile ability. This can be by means of pseudopodial extensions by which cells 'drag' themselves along. A second form of movement involves cytoplasmic streaming within the cell. Many developing systems rely extensively on cell movement, as described in Chapter 8.

PATTERN-FORMATION MECHANISMS

There are two fundamental mechanisms which utilize all the developmental parameters in one form or another to produce shapes and patterns in the developing organism. Firstly, by *pattern specification*, which is the way in which cells receive the specific information which they need to differentiate. Pattern specification normally takes the form of an interaction between external signals and the inherent genetic information in the cells. *Morphogenesis*, on the other hand, is the process by which cells differentiate using the specified pattern to form new shapes and structures during development. How might pattern specification work?

Imagine a one-dimensional row of cells. Suppose (and note that we are only hypothesizing here) that these cells all have the ability to form a pigment of one of three different colours, red, white, or blue (that is, the cells may enter one of three *differentiated states*). We want to generate three equally sized regions, one of red cells, one of white cells, and one of blue cells (in other words, a 'French flag'). How could this be done?

The French flag

A 'source' of production of some chemical substance might be set up. Such a chemical is termed a *morphogen*, a word derived from classical terms for 'shape' and 'generation'. If the morphogen source is at one edge of the field of the cells, it will diffuse across the field at a rate depending on the molecular weight of its constituent molecules. At the opposite side of the field a 'sink' is hypothesized. This destroys morphogen molecules at a constant rate, so producing a steady concentration drop of morphogen across the field. (If the sink were absent there may be a danger of morphogen concentration building up to a steady state all over the field.)

If we represent the fall in concentration graphically, we see that each cell in the row has a particular morphogen concentration (Figure 3.2). To produce regions of red, white, and blue, cells might simply be pre-

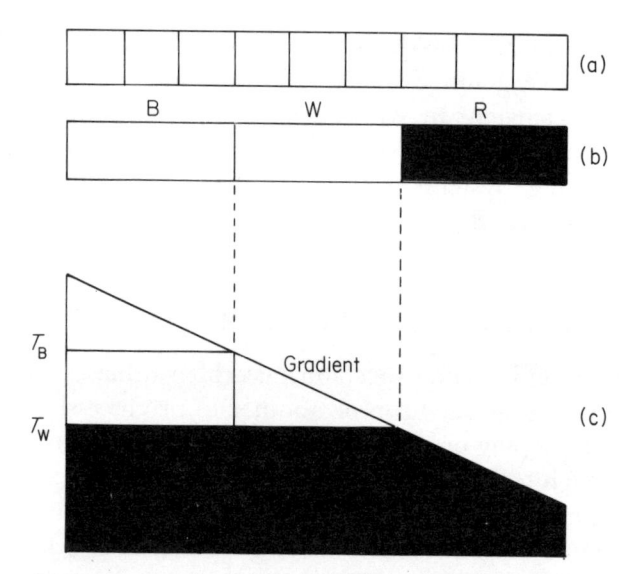

Figure 3.2 The 'French flag' model. A line of cells (a) undergoes differentiation so that one-third are blue, one-third white, and one-third red as shown in (b). This pattern could be specified by different levels of a morphogen so that if the concentration is above T_B the cells differentiate into blue cells. If the level is below T_W they differentiate into red cells, and if the level is between T_B and T_W the cells differentiate as white.

programmed with a specific instruction: for example, if the concentration of morphogen is between T_B and T_W then the 'differentiated state' or colour will be white. This sort of instruction might not be difficult to operate in practice, for example a cellular enzyme may be active only between a certain range of morphogen concentration.

Consider what happens if a gradient of the type outlined above is 'cut through' and the two halves are separated. Two types of regenerative growth can be considered. In the first type, *regeneration by remodelling*, there is no growth during regeneration, but missing parts are remodelled from the remaining stump: the regeneration of *Hydra* (a freshwater polyp) is an example. In this case, the regenerated system is complete but smaller than the original. Alternatively, regeneration in some developing systems is accomplished by *growth* of the cut stump, an

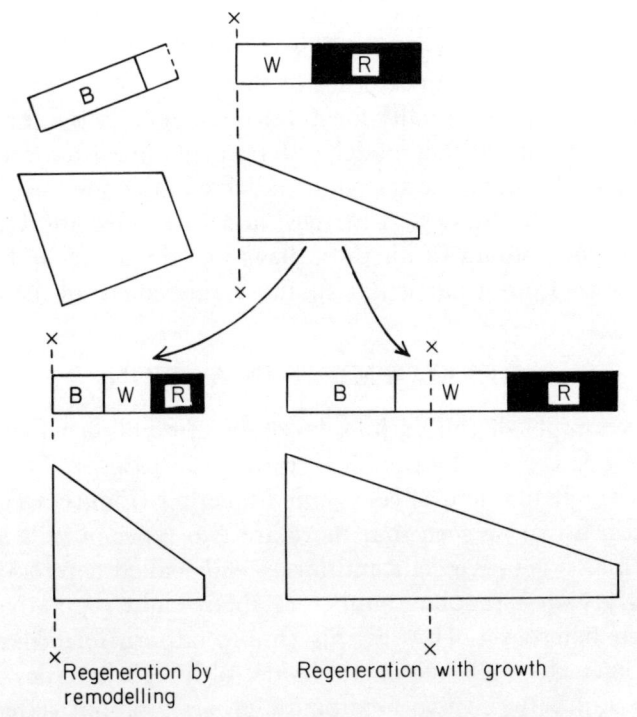

Figure 3.3 Regenerative growth and the French flag. If the pattern is cut at × ... × and the left-hand portion is removed, there are two possible outcomes. In regeneration by *remodelling*, missing parts are remodelled. Regeneration by *growth* involves cell division to produce the absent structures.

example of this being vertebrate limb regeneration. In this type of regeneration, cells at the cut surface are stimulated to undergo extra cell divisions. Consider what happens in each case in terms of our 'French flag' model (Figure 3.3).

If the blue region and part of the white region are removed, regeneration by remodelling results in the remaining white and red regions regulating to form the total 'French flag'. The cut surface becomes the new boundary of the field, and the gradient is reset so that the threshold limits of blue, white, and red are reset. In regeneration by growth, on the other hand, the gradient value at the cut remains unchanged, and new cells are generated until the full field size has been reformed. The lower gradient values then re-establish themselves in this regenerated region.

There is one especially attractive feature of gradients as specifiers of information which cells use to interpret their position in fields during development. This is that many diverse sorts of pattern could be generated from a very simple type of gradient. It is the *cells' interpretation of the gradient* which sets up the pattern, not the profile of the gradient itself. Lewis Wolpert, who originally formulated the idea of the 'French flag', proposed that using the flag model, different rules for interpretation with the same type of coordinate system could give in the one case the French flag, in another the Stars and Stripes, and in a third the Union Jack. Notice that the colours in all these flags are the same, but that each pattern derives from a particular spatial arrangement of the colours.

DEVELOPMENT IN ACTION

A simple example of the various levels of interaction which result in morphogenesis is given by the blue-green alga *Anabaena*. This organism exists as a single filament of cells joined together (Figure 3.4). On close examination it can be seen that there are two types of cells in the filament, normal vegetative cells and larger cells called heterocysts, which occur singly with a regular number of about eight vegetative cells between each heterocyst. How is this simple pattern formed and maintained? Vegetative cells regularly divide, although heterocysts do not. When the gap between two heterocysts grows to about sixteen cells, a new heterocyst forms from one of the vegetative cells *exactly in the centre* of the space between the two flanking heterocysts.

A simple model has been proposed by Wilcox and co-workers which uses the parameters of cell division, genetic information, environmental sensing, and differentiation. In brief, heterocysts are said to produce an inhibitory substance (signal) which travels outwards from each hetero-

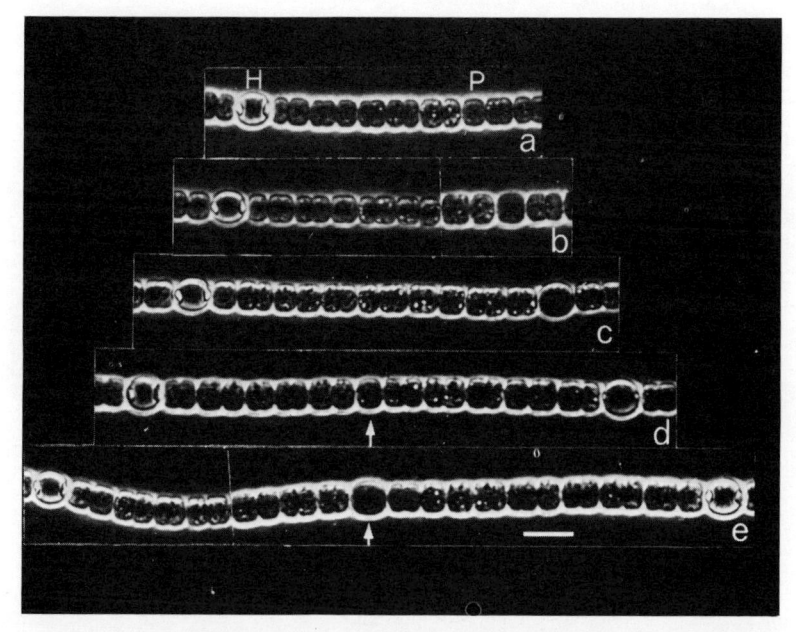

Figure 3.4 Growth of the *Anabaena* filament showing formation of pattern elements. H = heterocyst, P = proheterocyst. Arrow shows formation of new heterocyst.

cyst along the filament. As it does so, it may be absorbed or destroyed by surrounding vegetative cells giving a gradient of inhibitor concentration with a high point at each heterocyst. The concentration of the inhibitor substance would be lowest midway between each pair of heterocysts. There may be a threshold inhibitor concentration below which a vegetative cell can respond to the 'environmental input' and differentiate into a heterocyst. Figure 3.5 shows how the model might work.

There is an additional problem associated with the gradient, because it would be difficult to tune it so finely that only one cell receives the 'sub-threshold' level of inhibitor. *Anabaena* has a solution. Cell division occurs asymmetrically: each cell gives rise to two daughter cells—a long one a short one. Heterocysts can only develop from the short cells, thus cutting down the number of possible heterocyst precursors. A second possible mechanism is that two newly formed heterocysts (or proheterocysts) can inhibit one another; only the one producing most inhibitor will not revert to a vegetative cell.

In abstract terms, *Anabaena* shows many of the developmental parameters previously listed. Genetic information within each cell which can

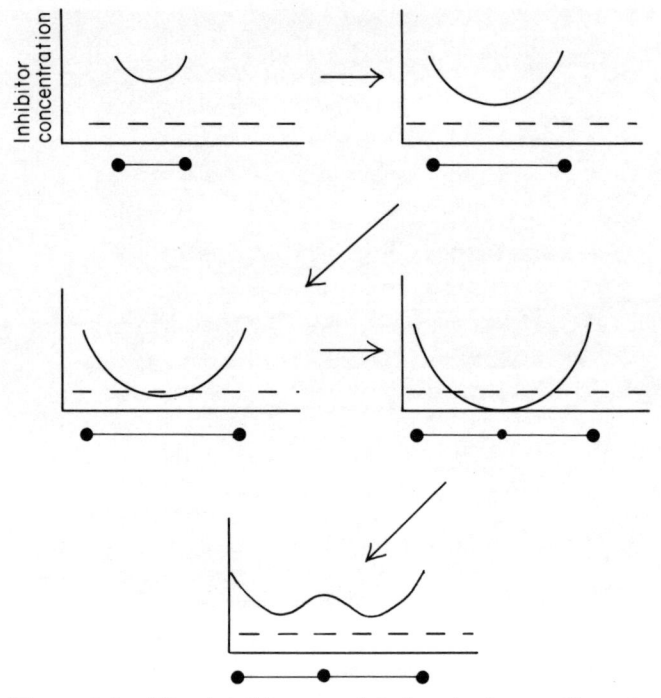

Figure 3.5 The inhibitor model for *Anabaena*. Growth of the filament (shown below each graph) results in a central lowering of inhibitor concentration between heterocysts. When the inhibitor level is lower than a threshold (dotted line) a new heterocyst may form. The new heterocyst also produces inhibitor.

control differentiation into heterocysts, vegetative cells, and the two types of precursor cells. Also present are an environmental sensing mechanism and cell division. Only cell movement is absent.

COMPLEXITY IN DEVELOPMENT

In the present chapter we have so far looked at the component processes of development. The *Anabaena* system hardly seems complex enough to model dynamically, although several attempts have been made (see Chapter 8). It is a big jump from a one-dimensional filamentous alga to an animal capable of many varied activities, so are there better examples which can illustrate the basic idea of how cells cooperate to build complex shapes and patterns in development?

One such example is given by the behaviour of cells growing *in vitro*.

Consider a layer of animal cells growing in a flat plastic dish. What sort of pattern will they take up? Elsdale (1970) studied this problem, finding that the cells often become arranged into parallel bundles (Figure 3.6). He noticed by using time-lapse cinematography that during bundle formation the cells undergo a kind of see-sawing motion together. Elsdale made the novel suggestion that the fibroblast bundles were behaving as

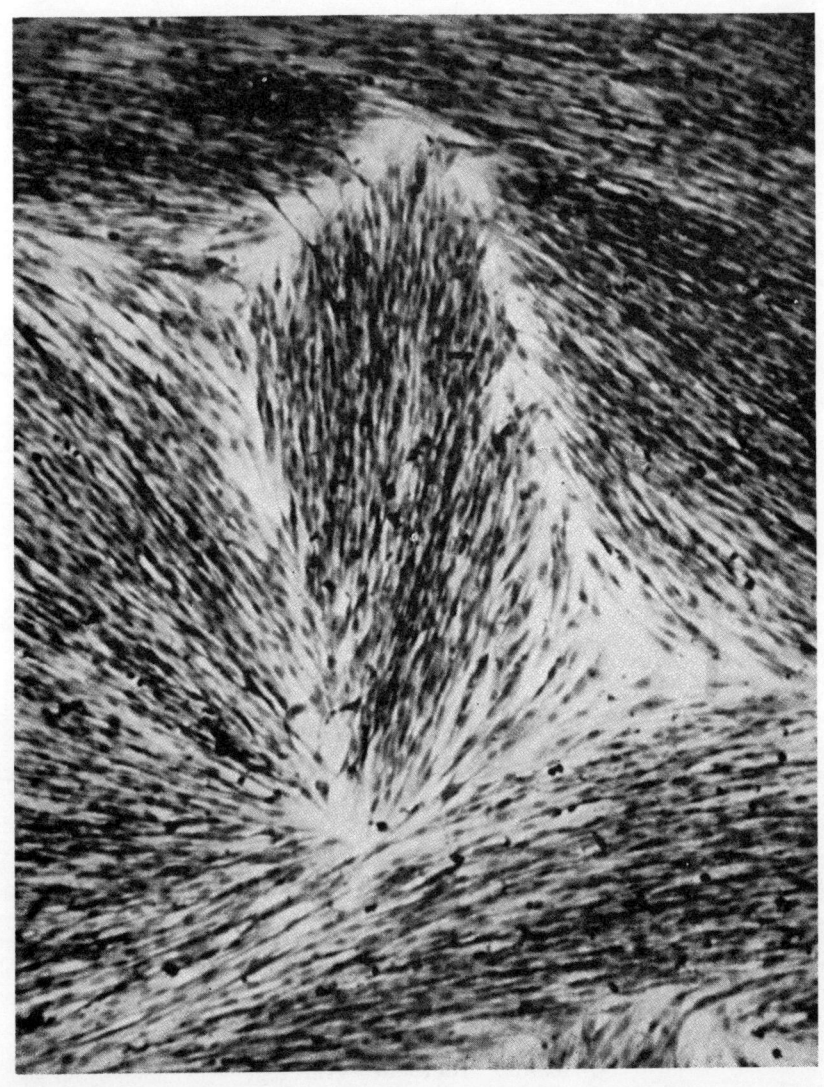

Figure 3.6 Fibroblast cell patterns in tissue culture. (Photograph courtesy of Tom Elsdale.)

an example of what is called in engineering an 'inherently precise' machine. One such machine grinds spherical lenses by randomly rubbing two blanks over one another, as shown in Figure 3.7.

Elsdale suggested that similar random movements could be acting in the case of the fibroblast bundles, only instead of forming a 'spherical' shape, the shape of the cells forces them into parallel bundles. This mechanism may be a simple illustration of the idea that large groups of

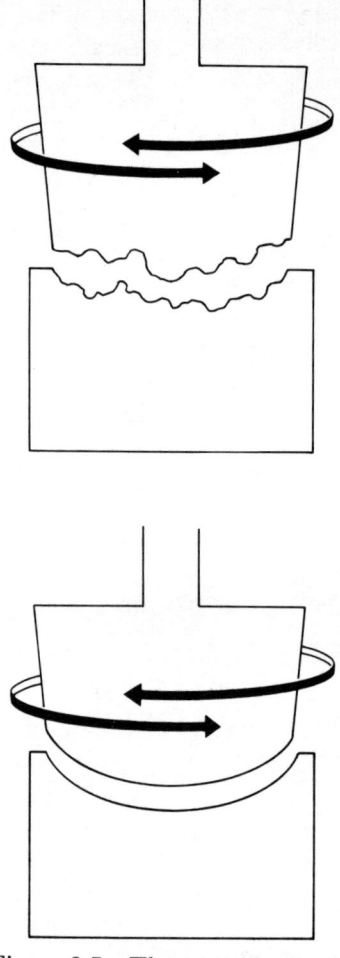

Figure 3.7 The working of an 'inherently precise' machine. Random grinding of rough glass lens surface produces a spherical lens.

cells, all doing identical simple things, can undergo a primitive form of morphogenesis. How could we tell if Elsdale's ideas are correct? The simplest solution would be to attempt to experiment biologically by interfering with the 'see-sawing' movement of the cells themselves. This procedure is fraught with difficulties, because arresting cell movement might have all kinds of side effects: the results would therefore be ambiguous.

An alternative technique would be to devise computer simulations to find out if simple rules modelling the cell movements can produce computer analogues of the fibroblast bundles. Whether or not this would work is an unanswered question, although Chapter 8 discusses other examples which do show that complex shapes may be generated by the repetitive application of simple rules of cell interaction.

The Elsdale type of cell-patterning mechanism is not the only simple example of cells interacting to form a relatively complex pattern. In some cases, differences in shape and patterns can be tracked down to a single gene difference. Changes in one particular genetic signal can produce, for example, smooth or crenulated fungal colonies. There are countless other cases, in organisms ranging from bacteria to man, of tiny differences resulting in gross changes in morphology. How the genes control development is more the subject of a book on developmental genetics, but the evidence of genetics certainly supports the hypothesis that 'simple' instructions can generate complex end-products, at least when cell interactions are being considered.

The classical study of embryology has centred around the early development of animals. One particular step in this development has been analysed in detail; this is gastrulation, the formation of the gut. The first stage in the development of an organism like a sea urchin, or even a frog, is the formation of a hollow ball of cells from the fertilized egg. This ball of cells is called a *blastula*, and could be likened to a hollow rubber ball. The gut forms by a process called *invagination*. One side of the blastula caves in, in much the same way as a rubber ball does if pushed with a finger. How do the cells accomplish this feat? It is an important step for the growing embryo, for eventually the innermost end of the cavity will join up with the opposite side of the embryo and will open to form the mouth aperture. The steps of sea urchin gastrulation are shown in Figure 3.8.

Tryggve Gustafson and Lewis Wolpert suggested that the events of sea urchin gastrulation could be accounted for by two cellular properties, changes in cell adhesiveness and cell shape. Although a consideration of their work takes us a little close to traditional developmental

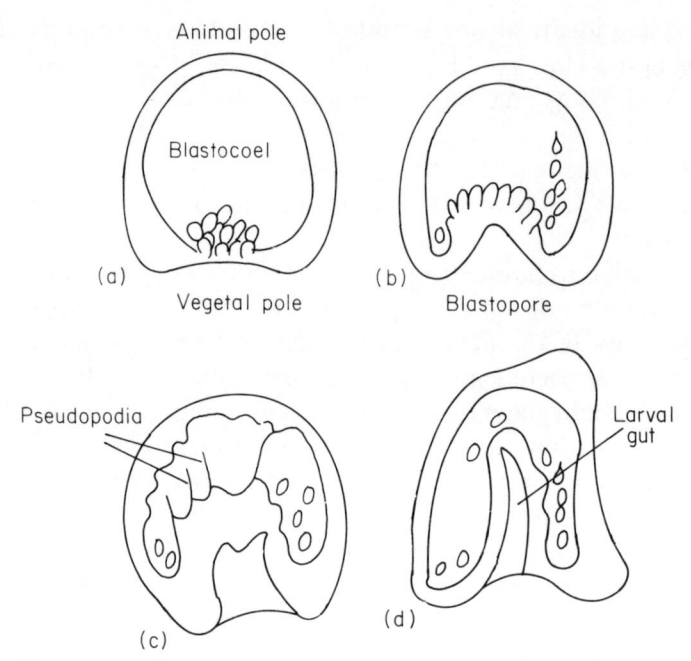

Figure 3.8 Gastrulation in the sea urchin shows how interaction between cells produces complex patterns during development. (a) The early blastula consists of a single cell layer around the central blastocoel. Cells migrate into the blastocoel from the vegetal pole (b) and invagination occurs forming the blastopore. As invagination occurs, pseudopodia anchor themselves to the animal pole region (c) and the larval gut is formed (d).

biology, it represents one of the few closely examined examples of how cells interact during development and as this is such a central element of the type of models discussed later, no apology is made.

Gustafson and Wolpert's argument was as follows. Consider an idealized cell in contact with a flat base (Figure 3.9). If there is little adhesion between the cell and the underlying base then the cell will not have much contact with the base, and (at least in the case of our idealized cell) will be rounded in shape. If the amount of adhesion increases, then there will be more contact between the cell and the base, and the cell will become flatter by stretching as shown in Figure 3.9(b). Hence the extent of contact between a cell and its base (or perhaps between two cells) depends on a balance between the forces that tend to increase mutual contact (for example, increased adhesiveness) and the forces that resist deformation.

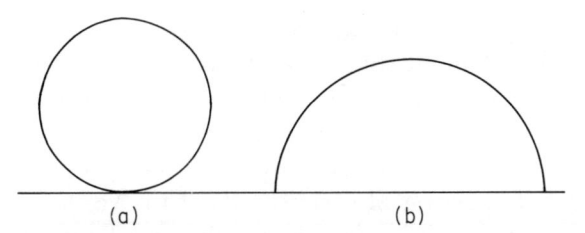

Figure 3.9 Changes in cell shape accompanying variation in cell adhesiveness.

If we take the argument one step up from single cells to sheets of cells, can we produce specific tissue shapes by this mechanism? In the sea urchin, the cells at the part at which the invagination to form the embryonic gut starts are attached to a membrane on their outer edge. If we consider that the adhesion between the cells is moderate, a diagrammatic representation of the cells and membrane might look like Figure 3.10(a). Changes in cell adhesiveness lead to changes in shape; for example, if the cells become more adhesive to both themselves and to the membrane then the resulting situation would be as shown in Figure 3.10(b). If the cells became 'stickier' only to themselves, then the situation illustrated in Figure 3.10(c) would result.

Gustafson and Wolpert proposed that the cells at the vegetal pole resembled those in Figure 3.10(b) before invagination. They suggested that contact between adjacent cells is then reduced, but contact between the cells and the hyaline layer remains the same. If the ends of the cell

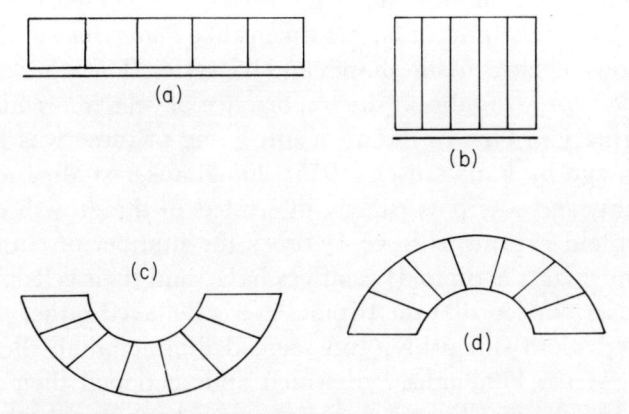

Figure 3.10 The effect of adhesiveness changes in the conformation of a sheet of cells.

sheet are fixed (by some characteristic of the wall of the invaginating blastula), the sheet cannot spread, so it curves. This situation is shown in Figure 3.10(d). The curve this time is *inwards*, forming the archenteron, a cavity which later forms the gut. Gustafson and Wolpert's model therefore suggests that 'primary invagination' could result from a change in the adhesiveness of certain cells at the vegetal pole.

Turning now to consideration of the second of the two properties, changes in cell *shape*, we can look at what happens when the primary invagination has got under way. The second phase of gastrulation is marked by the formation of *pseudopodia* by the cells at the archenteron tip. These cells seem to have a stretching function, the pseudopodia anchoring themselves to the wall of the blastula around the animal pole (see Figure 3.8) and then contracting, and so mechanically helping the elongation of the invaginating gut. This process seems to be very important, for if the gut cavity is treated with sucrose solution, producing a change in pressure inside the cells, the pseudopodia break down and archenteron invagination stops.

With the accent on changes in cell adhesiveness during the primary invagination of the gut, and cell shape changes with pseudopodial formation during secondary invagination, Gustafson and Wolpert underlined the importance of simple cell properties in seemingly complex morphogenetic events.

THE BUILDING BLOCKS OF DEVELOPMENT

The author's bias certainly tends towards the view that cells are the important building blocks of development. It would be unfair to conclude this chapter without reviewing a little of the evidence against this viewpoint. Certainly unicellular organisms like *Paramecium* and amoebae have their own characteristic shapes and life styles. There are also experiments which throw doubt on the importance of cells in certain types of pattern formation. One of the most intriguing of these was performed some years ago by Fankhauser (1955). Fankhauser studied amphibian development, and was particularly interested in the growth of triploid newts. Triploid organisms have $1\frac{1}{2}$ times the number of chromosomes that their normal (i.e. diploid) relatives have, and their cells are bigger.

Fankhauser noticed that the triploid newts behaved rather peculiarly. They moved more sluggishly, and seemed dimmer than their diploid colleagues. When Fankhauser dissected and sectioned their brains he found an interesting difference which may have accounted for the behavioural differences. The brains were the *same size*, but the triploid newts of course had fewer brain cells because of the larger size of the cells. Hence the triploid brains had fewer neurons. The most interesting conclusion from this experiment is that brain size is dependent not on cell *number* but on the overall, non cell-related distance between opposite sides of the brain, a parameter more likely to be associated with some kind of biochemical diffusion field than with cell number.

The controversial issue of the relative importance of cell interactions and biochemical gradients in specifying developmental patterns might well be resolved by the construction of the right kind of model of the right developmental system. In the following chapters we go on to look in detail at the whole range of developmental models so far constructed.

REFERENCES

This chapter has done little more than indicate the important components and problems associated with animal development. Several excellent textbooks on developmental biology have been published in recent years, and these are listed in Appendix 2. Further reference to the developmental systems described above may be found in the following publications.

Elsdale, T. (1970) Pattern formation and homeostasis, *Ciba Foundation Symposium on Homeostatic Regulators* (G. E. W. Wolstenholme and J. Knight, Eds.), Churchill, London.

Fankhauser, G. (1955) in *Analysis of Development* (B. H. Willier, Ed.), Saunders, N.Y.

Gustafson, T. V. (1969) Cell recognition and cell contacts during sea-urchin development, in: *Cellular Recognition* (R. T. Smith and R. A. Good, Eds.), Appleton-Century Crofts, N.Y.

Wilcox, M., Mitchison, G. J., and Smith, R. J. (1973) Pattern formation in the blue-green alga *Anabaena*, *J. Cell Sci.*, **12**, 707–723.

Wolpert, L. (1969) Positional information and pattern, *Journal of Theoretical Biology*, **25**, 1–49.

Chapter 4

Substitute system models

Any journal featuring articles or papers on developmental biology will be full of examples of 'substitute system' models. They are so common that it is sometimes difficult to know what process they are models of, and indeed we are surely all guilty of 'model chauvinism' to one degree or another. Does a mouse geneticist understand the relevance of the work of his bacterial colleagues? Does the amphibian embryologist pay heed to the findings of the slime-mould biologists?

The problem does not lie merely in the choice of individual system, or elements of a system, which a particular biologist has chosen a substitute system to study. More important is the dilemma of relevance of the substitute system: does it really show the important features of the general process that the experimenter wishes to study? A second factor is also important. Different substitute systems involve different degrees of handling facility, and this must also be taken into account by the researcher.

THE RELEVANCE OF THE SUBSTITUTE SYSTEM

The previous chapter considered the important parameters of developmental processes, and it is possible to use these as an index of the kinds of abstract processes which substitute systems are used to study in developmental biology. The most obvious process is development itself. All developmental biologists study the processes of growth and development, and would probably agree that the central goal would be to find out how an adult mammal grows from a single-celled egg. So far so good, but how can one assess the relevance of studies on organisms as diverse as bacteria or gorillas to human development in particular. It is necessary to be much more specific in problem definition.

There are two major problem areas within developmental biology which were described in Chapter 3: *differentiation* and *pattern specification and morphogenesis*. Differentiation involves state changes within cells, whilst pattern formation and morphogenesis are the processes by which spatial arrangements of state changes occur in cell populations and which give rise to biological shapes and patterns.

The non-spatial element of differentiation makes it a somewhat simpler problem to study than pattern development, and a variety of substitute system models have been used to model the way in which state changes in cells occur. Because the end-product is to analyse cell state changes in biochemical terms, a favourite organism for this kind of study has been the bacterium. A bacterium possesses a chromosome and genes, and produces particular biochemicals in response to stimuli. It therefore shows a simple form of 'differentiation'. By using the human intestinal bacterium *Escherichia coli*, Jacob and Monod (1961) were able to formulate a new hypothesis about how protein synthesis may be controlled in cells. Although there is still confusion about the applicability of this type of control system in higher organisms, there is no doubt that Jacob and Monod's work has inspired new insight into the control of gene expression. A more relevant 'differentiation' system in bacteria is the process of sporulation. Here, the single-celled bacterium actually shows cell-type differentiation—under certain conditions the bacterium may either divide to give two identical daughter cells, or it may become a spore, capable of surviving under conditions adverse to growth.

Not only whole organisms, but also parts of organisms can be used as substitute system models. The advantages of bacteria for studying differentiation are clear, but their simplicity makes the application of bacterial findings to higher organisms rather limited. This criticism cannot be levelled so easily at *in vitro* culture techniques. Here, cells from higher animals or plants may be grown in culture medium and their biochemical reactions studied. In some cases it is possible for spectacular forms of differentiation to occur. Steward found in the 1950s that single carrot cells could produce normal carrot plants after growth in culture (Steward *et al.*, 1966). Similarly, study of teratocarcinoma cells in culture can lead to further understanding of differentiation in cancerous tissues.

The formation of patterns involves study of a different range of substitute system models. Ideally, model systems with only two or three different cell types and a minimum of cells should be studied to see how the principles of pattern specification and morphogenesis work. Probably a major fault of developmental biologists is that they have failed to search for such simple systems and have distributed themselves amongst a plethora of complex and different systems. We are therefore in a kind of developmental 'Tower of Babel', with everybody speaking a different language with little common ground.

Escherichia coli is a single-celled organism and possesses no multicellular pattern. One of the simplest patterns to be studied in recent years is that seen in the blue-green alga *Anabaena*, discussed in the previous chapter. With only two types of cells arranged in a filament, it might seem that *Anabaena* could answer the basic problem of how cells communicate with one another to produce simple patterns. The answer is rather disappointing. We suspect that levels of a biochemical which diffuse along the filament can signal the cells to form the pattern, but there are immediate criticisms of *Anabaena* as a model system. It shows no cell movement, which is thought to be important in higher organisms. The cells are also cellulose-walled 'boxes' which keep rigidly to the single cell width filament, and are far from the amorphous-shaped cells seen in developing animal systems.

More complex substitute systems are often used to study pattern specification and morphogenesis, and the degree of complexity involved is often related to the 'dimensionality' of the model system. The *in vitro* behaviour of fibroblasts which form a simple two-dimensional pattern has been discussed earlier (p. 26), and the formation of two-dimensional sheets of larval insect cells that metamorphose to form adult structures will be discussed in detail in Chapter 9. Three-dimensional patterns are the most complex of all, and a common example studied by embryologists is early cleavage of the egg cell in a variety of organisms. In some cases it is possible to consider three-dimensional growth as a two-dimensional problem (the general problems inherent in this type of analysis are discussed in Chapter 7). The development of the vertebrate limb is often looked at in this way. Pattern specification and morphogenesis therefore present a particular difficulty in substitute sytem modelling. In bald terms, you 'get what you pay for'. The simpler the system, the less relevant it seems to be. It is partly because of this pessimistic situation that substitute systems have themselves been the subject of so much computer and mathematical modelling, to be described elsewhere in this book.

THE VARIETY OF SUBSTITUTE SYSTEMS

It is also important to consider how amenable the model itself is to analysis. Even different species of the same genus of fruitfly vary dramatically in their receptivity to a laboratory environment, and this type of problem accounts for the popularity of the 'laboratory favourites'. Developmental studies normally revolve around a dozen or so organisms which might be split into three 'relevance groups' (Table 4.1). The first

Table 4.1 Groups of organisms used to study developmental processes.

Group 1	Group 2	Group 3
Mammals, e.g. mouse, rat	Insects, e.g. *Drosophila*,	*Anabaena*
Fowl	*Rhodnius*,	Slime moulds
Amphibian	*Galleria*	Bacteria
	Molluscs, e.g. *Lymnaea*	
	Sea urchin	

group comprises vertebrates, studies of which may be directly relevant to mammalian development. Group two are the invertebrate animals, and animals of this class are the subject of much debate concerning relevance. How can insect metamorphosis be equated to mammalian organogenesis? How can growth of sea urchin tube feet be relevant to vertebrate limb growth? Group three organisms are the poor relations of the classification. They are often almost alien in character to vertebrate development (compare *Anabaena* and vertebrate limb patterning, for example). Nevertheless, the patterning process in these simple organisms may give insight into how more complex patterns can develop.

NON-BIOLOGICAL SUBSTITUTE SYSTEMS

This is an area that has not been studied in much detail by developmental biologists. Waddington (1962) described the morphogenesis of membrane systems in cells like chloroplasts and the light-sensitive receptor cells in insect eyes (Figure 4.1). In both cases Waddington suggested that a model system for the spontaneous generation of these membrane forms may be illustrated by the interaction of certain lipids like lecithin and cephalin with water. If a piece of lipid is placed on a microscope slide in a drop of water, it will gradually form patterns like those of Figure 4.2. Waddington's description of the morphogenesis which occurs is most graphic:

'Over the course of two or three days the preparation as a whole goes through a fairly definite life history. At each stage the forms, which are gradually increasing in water content as time passes, take up a number of relatively characteristic shapes . . . At first the forms protrude more or less radially from the central fragment. It is very

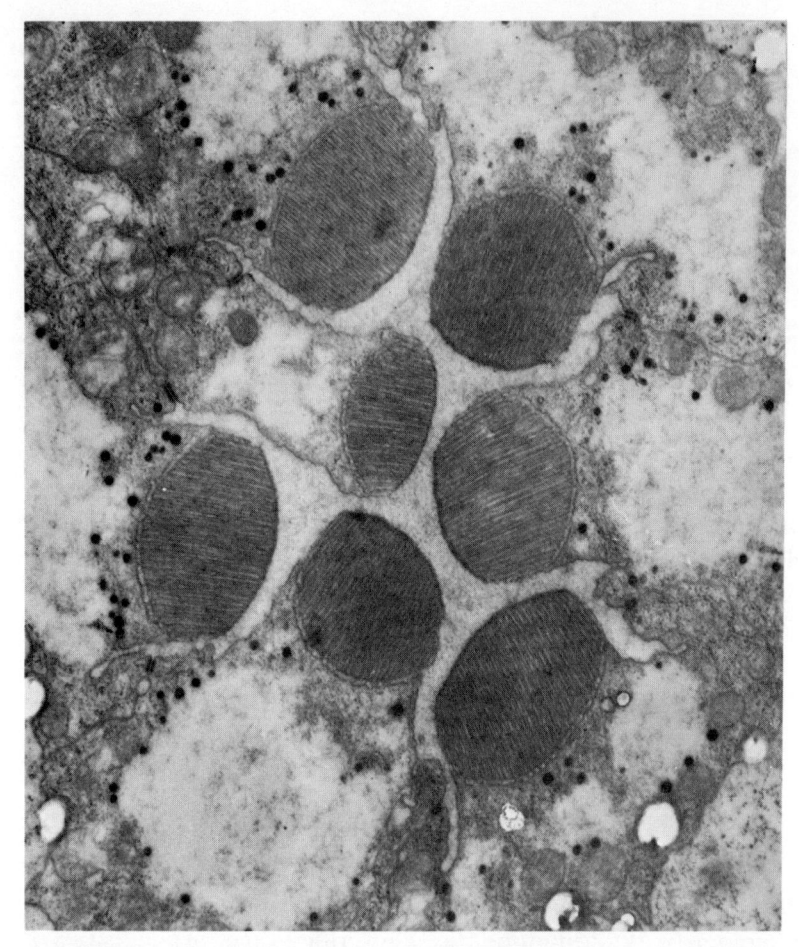

Figure 4.1 Membrane system in the *Drosophila* compound eye. Stacked microvilli contain the visual pigment. (Photograph courtesy of Andrew Tomlinson.)

striking to observe the complexity and relative regularity of the arrangements which they may exhibit. We may find, for instance, long stalks, each of which bears a roughly spherical head, consisting of a ball of twisted, tubular structure: other forms which are made up of a zigzag of swollen chambers of very regular size and arrangement, and also a number of other regular types. At a later stage, the elongating forms bend round and tend to run circumferentially round the mass . . . By this stage the lipid is becoming highly swollen with a high water content, and there are areas which show a regular arrangement of differences in refractive index, that is to say, of regions of close or distant packing of the individual lipid

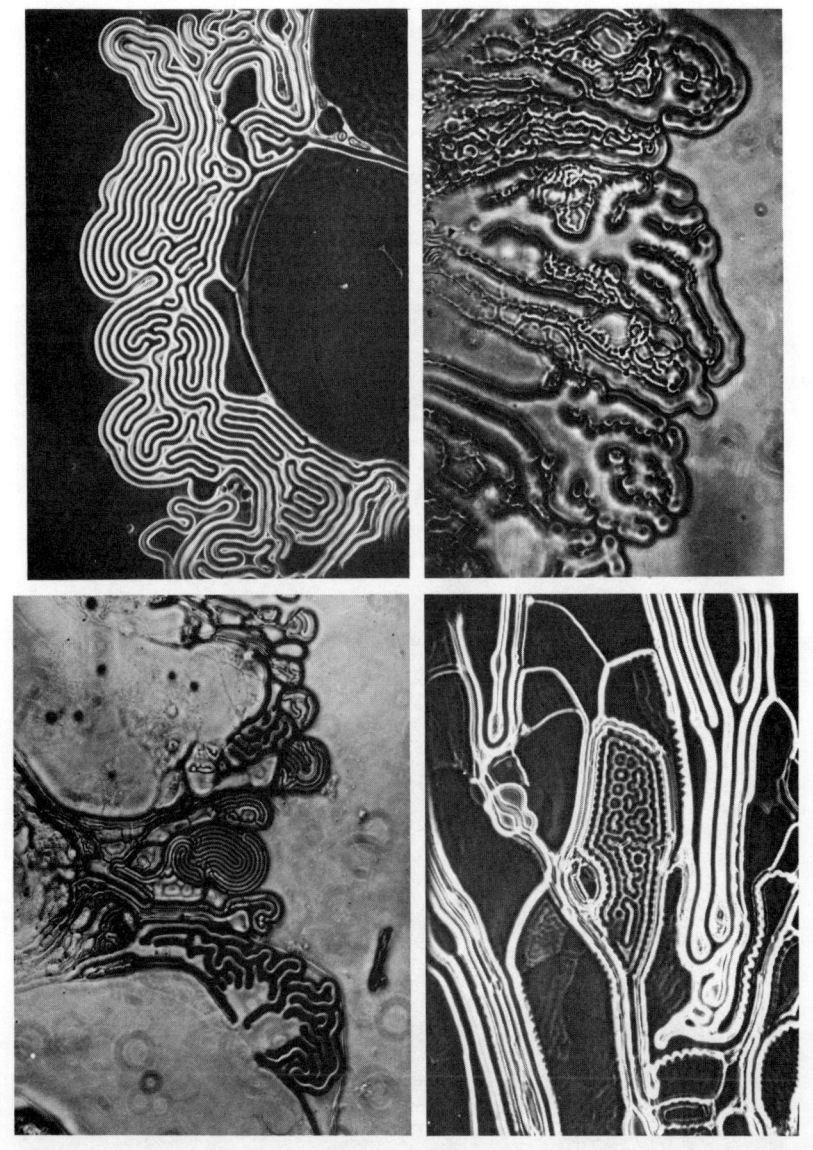

Figure 4.2 Myelin figures produced by the interaction of lecithin with water (by courtesy of Dr G. G. Selman).

sheets. These may take the form of hexagonal close packing, or various more complex periodic structures.'

The high degree of complex patterning produced by the simple molecular interactions between water and lipid is very striking and may well give clues as to the generation of biological membrane structures in living tissue.

Chemical patterns have also been studied as models of biological pattern formation by Winfree (1973), who analysed the forms produced by waves of chemical activity through a motionless liquid. If ferric ions are allowed to catalyse the oxidation of malonate by bromate in acid solution in a petri dish, complex oscillatory patterns are produced (Fig-

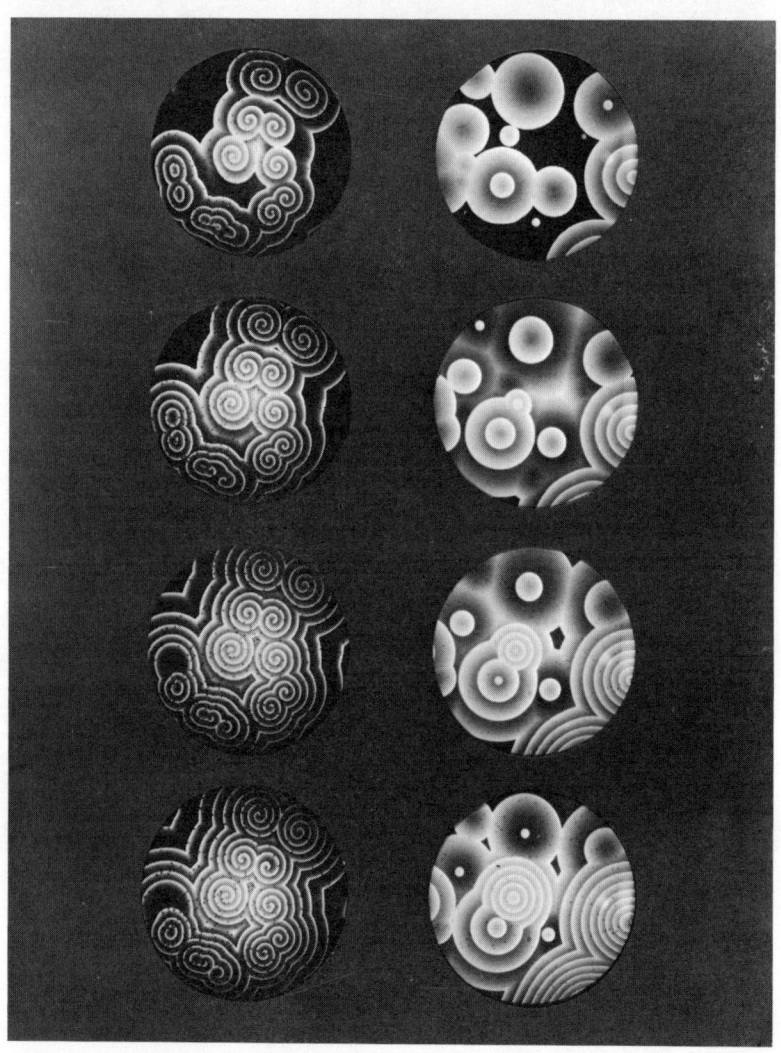

Figure 4.3 Chemical patterns produced by the oxidation of malonate by bromate in acid solution. (Photograph courtesy of A. T. Winfree.)

ure 4.3). Such demonstrations that complex patterns may be set up by using relatively simple chemical reactions suggest that biological patterns might be generated by similar means. The chemical reaction therefore represents a possible model of biological pattern formation.

A third example of a non-biological substitute system concerns the packing of cells in tissues. Cells normally have between five and seven neighbours when packed in two dimensions—what processes govern packing in living tissues? The answer is probably given by observation of the behaviour of liquid films such as soap bubbles *en masse*. Surface-tension considerations determine the optimum packing ratio amongst the bubbles, which pack in a similar manner to cells in living tissues, especially as both cells and bubbles have a tendency to 'round up' on isolation.

One of the most clever examples of a substitute system model for developmental biology was constructed by Lawrence (1970). Lawrence studied the cuticle of the milkweed bug *Oncopeltus*. The cuticle of this insect is segmented. On close examination of the segment borders, gaps were occasionally seen (Figure 4.4(a)). The pattern of cuticular hairs around such gaps was disturbed. Normally the hairs point in an anterior–posterior direction, and as each hair is made by an underlying epidermal cell, the hair orientation indicates the polarity of the cell. What mechanism controls this orientation? Locke had earlier proposed (1966) that insect segments have a gradient from one border to the other: the gradient serially repeats itself in each segment, and would be responsible for the pattern of the *Oncopeltus* hairs. The cells would simply align themselves along the line of the gradient.

Lawrence proposed that the segment gaps in *Oncopeltus* were evidence in favour of this model, and he set out to prove it by constructing a 'sand model' to represent the serially repeating gradient (Figure 4.4(b) and (c)). Glass plates represent the segment margins separating high- and low-gradient points. A stable sand slope between two glass plates represents the gradient. If a gap appears in a glass plate, sand flows from one segment to another, continuing until new stable slopes are established. When movement stops, the sand is aligned in the reverse direction: gradients with intermediate orientation appear in both segments. The maximum slope of these intermediate gradients fits well with the observed orientation of the *Oncopeltus* hairs. The results of this 'experiment' strongly suggest that the polarity of the epidermal cells depends on the orientation of the steepest gradient slope. As in the model, when an unstable situation develops as a result of sand at different gradient levels being placed together, the gap in the insect segment interferes with

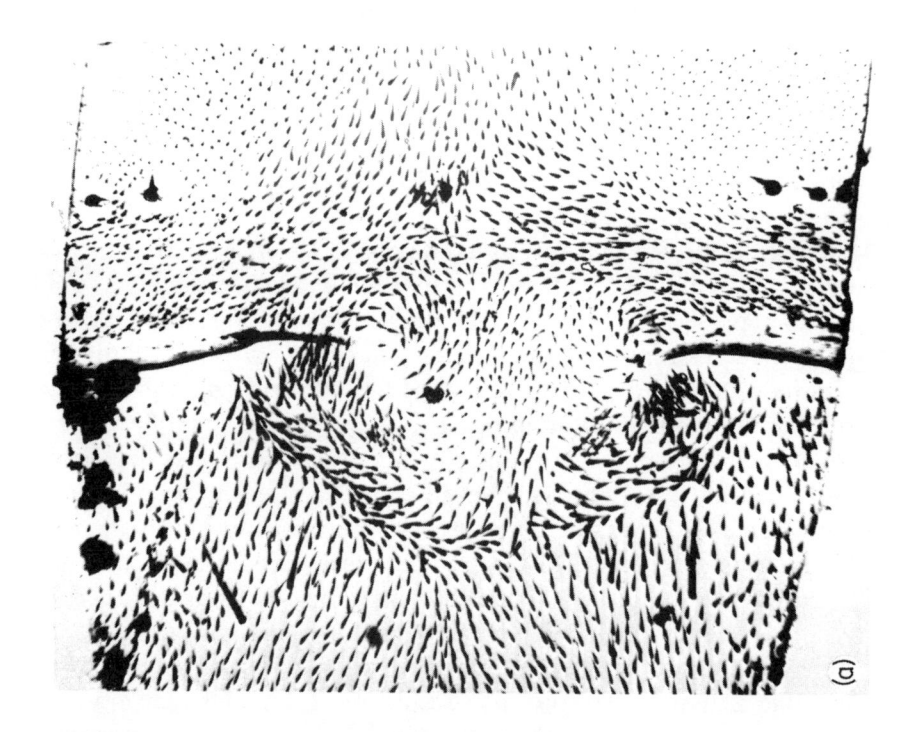

(b)

(a)

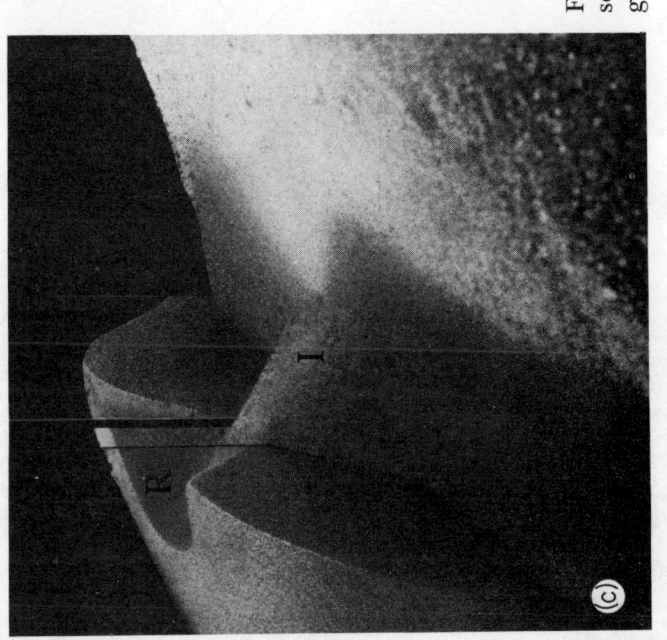

Figure 4.4 The gradient model in *Oncopeltus*. (a) Gap in the segment border. (b) Sand gradient. (c) Gap in the sand gradient. (Photographs courtesy of P. A. Lawrence.)

the normal pattern of the gradient, so producing an abnormal polarity pattern. The problem of insect cuticle patterns will be further discussed in Chapter 8, where computer models generating such patterns will be described.

Gmitro and Scriven (1966) considered several physical analogues of biological pattern formation. One of these occurs in a shallow dish of ordinary liquid that is being uniformly heated over its bottom surface. A regular hexagonal tesselation of so-called Bénard cells appears as shown

Figure 4.5 Development of Bénard cells.

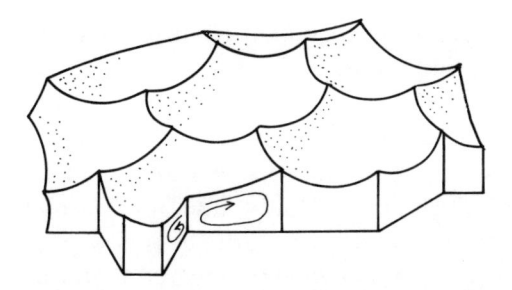

Figure 4.6 Heat flow within an individual Bénard cell.

in Figure 4.5. The cells are formed as follows. Hotter, more buoyant fluid rises from the bottom of the dish and the colder fluid at the surface sinks. Hot and cold columns therefore exist side by side throughout the fluid. There are also lateral forces and temperature gradients opposing this flow. The presence of the side of the dish favours concentric ringed cells at first, but these settle down to give the hexagonal arrangement of Bénard cells, provided the dish diameter is considerably greater than the natural cell size. The flow within a single cell is shown in Figure 4.6. The cells remain fixed in location and the flow within them is steady.

Biological patterns are also often hexagonal. Apart from the obvious example of cell packing, the hexagonal arrangement of the insect eye could also be considered. This latter example does bear certain superficial similarities to the Bénard cell pattern. The insect eye (see Chapter 9 for fuller details) consists of a hexagonal arrangement of individual 'eyes' called ommatidia. Each ommatidium consists of around twenty individual cells. The basis of the formation of the hexagonal pattern is unknown, but the presence of local 'activity' centres might perhaps force the cells into hexagonal structures in a similar manner to the way in which Bénard cells are formed. The mathematical basis of Gmitro and Scriven's work is considered in the next chapter.

A second hexagonal patterning mechanism is considered by the same authors, and occurs when solids are frozen out of molten solution. There is a tendency for solute to be redistributed at the freezing face of the solid. It is caused by local supercooling and is opposed by thermal transport and reduced by diffusional transport. Gmitro and Scriven make the point that *the static structure is merely a partial record of the dynamic processes by which it is produced*, rather than being an end in itself. They abstract the important factors of pattern formation as being threefold: (1) transformation processes; (2) transport processes; and (3) coupling of the two types of processes together. Their mathematical formulations begin from this starting point.

We have looked at a variety of substitute system models, some of which involve simpler biological organisms, whilst others involve substitute physical systems. There is no doubt that models of these types will continue to be of importance in the future. As new practical techniques became available, a new range of experimental material will be brought into the group of 'useful' organisms, at least in developmental terms. We have seen in recent years the resurgence of interest in microorganisms, for example, as genetic engineering techniques have allowed isolation and analysis of individual genes. Conversely, the organisms studied today may be of little importance tomorrow. Much of the emphasis of early embryological work was placed on obscure insects which are now no longer studied, precisely because *Drosophila* turned out to have such great genetic potential and is so easy to rear. Table 4.1 could be obsolete in twenty years and there is no way at present to predict its future form.

REFERENCES

Nearly all research in developmental biology involves substitute systems to one degree or another. It is therefore difficult to recommend particular examples of the use of this type of model. Pattern specification and morphogenesis are dealt with regularly in the journals *Roux's Archives of Developmental Biology, Journal of Embryology and Experimental Morphology*, and *Developmental Biology*. *Developmental Biology* also contains many examples of models of differentiation, together with the journals *Cell* and *Differentiation*. Other references quoted in this chapter are:

Gmitro, J. I., and Scriven, L. E. (1966) A physicochemical basis for pattern and rhythm, in: *Intracellular Transport*, Academic Press, N.Y.

Jacob, F., and Monod, J. (1961) Genetic regulatory mechanisms in the synthesis of proteins, *J. Mol. Biol.*, **3**, 318–356.

Lawrence, P. A. (1970) Polarity and patterns in the postembryonic development of insects, *Adv. Insect. Phys.*, **7**, 197–260.

Steward, F. C., Kent, A. E., and Mapes, M. O. (1966) The culture of free plant cells and its significance for embryology and morphogenesis. In *Current Topics in Developmental Biology* (Eds. A. A. Moscona and A. Monroy), **1**, Academic Press, N.Y.

Waddington, C. H. (1962) *New Patterns in Genetics and Development*, Columbia University Press, N.Y.

Winfree, A. (1973) Scroll-shaped waves of chemical activity in three dimensions, *Science*, **181**, 937–938.

Chapter 5

Mathematical models of development

INTRODUCTORY NOTE

There is no clear dividing line between computer model and classical mathematical model but rather a continuum which is spanned by few biologists. At one extreme is the computer model founded in logical manipulations, with little mathematical basis save that of, perhaps, description in terms of symbolic logic. At the other extreme lies the biologists' nightmare: the model couched in terms of calculus or topology. Even D'Arcy Thompson, the father of the mathematical treatment of development, wrote in 1917 in the preface to his epic *On Growth and Form*:

> 'It is not the biologist with an inkling of mathematics, but the skilled and learned mathematician who must ultimately deal with such problems as are sketched and adumbrated here ... the mathematical methods which I have introduced are of the easiest and simplest kind.'

It was therefore with trepidation that the present biologist approached this chapter. Mathematicians tend to write for their own kind: even the enlightened ones that write about development. I hope the mathematical reader will forgive the simplistic interpretation, and that the biologist will revel in this same simplicity.

THE SPECTRUM OF MATHEMATICAL MODELS

Mathematical models are taken here to mean models that involve algebraic, symbolic, or arithmetical manipulations. The distinction between models of this type and computer models proper is, as stated above, rather arbitrary, and several of the models introduced in Chapter 8 could be considered equally well in the present chapter. For the present purposes, the separation has been chosen in the following manner. Models involving a significant proportion of simulation have been dealt with under the heading of computer models. Those that are either in abstract mathematical form, or have not been simulated on the computer to a significant extent, are considered in the present chapter.

Mathematical models of development have been split into two categories. Firstly, those involving classical mathematics: calculus and algebra, concentrating on spatial models of pattern formation. The second group of models to be analysed involve the use of catastrophe theory, a branch of topology which has been extensively used to describe developmental processes in recent years.

Mathematical models are the 'purest' form of model because every individual parameter and rule is so precisely defined. There are nevertheless considerable problems when it comes to both applicability and relevance. Some mathematical models are so vague that it is difficult to see how they can ever help us to understand real processes except in the most obscure way. It is difficult for a non-mathematician to avoid the trap of criticizing what he does not understand. However, it is equally difficult to avoid the 'emperor's new clothes' syndrome. The truth is going to be somewhere in between the two extremes.

MATHEMATICAL MODELS OF PATTERN FORMATION

Thompson's contribution to biology was largely descriptive, albeit in a mathematical manner. His interest to developmental biologists lies in this descriptive approach. One example is given by Thompson's consideration of leaf and cone arrangement in plants. Thompson was the first to acknowledge his debt to his distinguished forebears:

> 'Leonardo da Vinci would seem . . . to have been the first to record his thoughts upon the subject, but the old Greek and Egyptian geometers are not likely to have left unstudied or unobserved the spiral traces of the leaves upon a palm stem, or the spiral order of the petals of a lotus or the florets in a sunflower. For so, as old Nehemiah Grew (1682) says, "from the contemplation of plants, men might first be invited to Mathematical Enquiries".'

Anyone who counts the spirals on the head of a sunflower or on a pine cone will discover that their number is usually a terms of the series 1, 1, 2, 3, 5, 8, 13, 21, ... This is the Fibonacci series, each of whose terms is the sum of the preceding two. Although Thompson describes the occurrence and structure of these natural patterns at length, he wrongly concluded that the Fibonacci series is an inevitable property of a regular leaf arrangement, and it was not until 1977 that Mitchison proposed a model to explain the natural Fibonacci series in plants.

Thompson's 'soap bubble' model for cell contacts has already been mentioned in the previous chapter. He also considered the problem of cell division—especially in situations where cellular bisections occur to produce particular shapes and patterns during development. By simple geometry, Thompson indicated that cell divisions normally occur by means of a partition of minimal area and that the following rules should hold: (1) the partition must cut across the longest axis of the figure, and (2) each partition must run at right angles to its immediate predecessor, meaning that (3) the dividing wall of the daughter cells is such that its area is the least possible by which the given space-content can be enclosed.

Thompson gave several geometrical examples to illustrate these points, going on to show how real organisms also fit the hypothesis of 'areae minimae'. Algae show the principles simply. *Spirogyra* is a thin cylindrical filament (Figure 5.1): here the partition lies transversely to the long axis of the thread. The flattened disc-like alga *Erythrotrichia discigera* (Figure 5.2) appears more complex, but Thompson showed it to obey the rules of cell division outlined above. The first two cell divisions are clearly the shortest bisecting partitions of a circle. The later divisions are within the four quandrants produced by these two divisions. Thompson went on to show that the bisection of the quadrants in geometrical terms to give least area should be *anticlinal* (cutting one radial wall and the periphery). This is the case in *Erythrotrichia*.

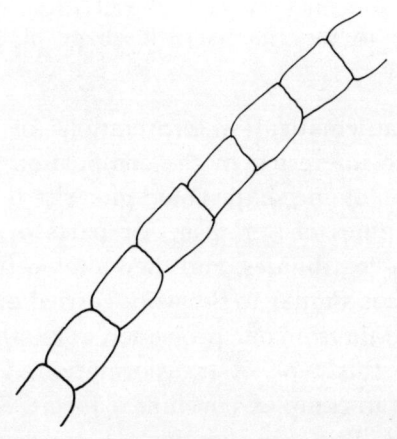

Figure 5.1 Segmental appearance of a row of *Spirogyra* cells.

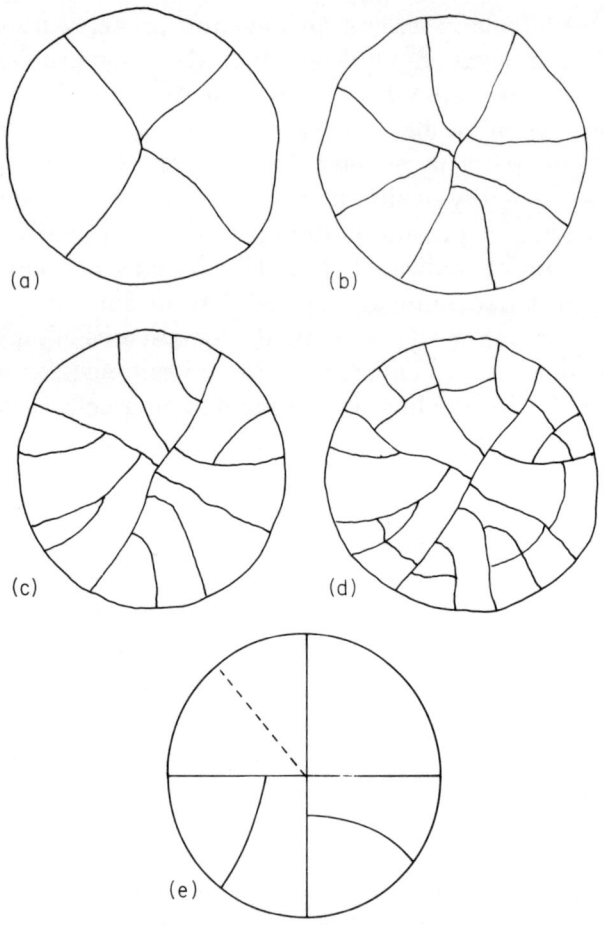

Figure 5.2 (a)–(d) Development of *Erythrotrichia*. (e) The anticlinal method of cell division which gives rise to the characteristic shape of the cells in the alga.

Besides the mathematical interpretation of biological shapes, Thompson was also interested in the comparison of related forms, or how the deformation of one shape could give rise to another. In general terms, he drew outlines of organisms or parts of organisms within a system of Cartesian coordinates, and then altered the dimensions of the axes. In a sense this is similar to the work carried out by a cartographer who transfers map data from one projection to another. Thompson gives many examples of this type of transformation, varying from simple linear deformations to complex curvilinear forms. Several examples are shown in Figure 5.3. The correspondence between the cannon-bones of

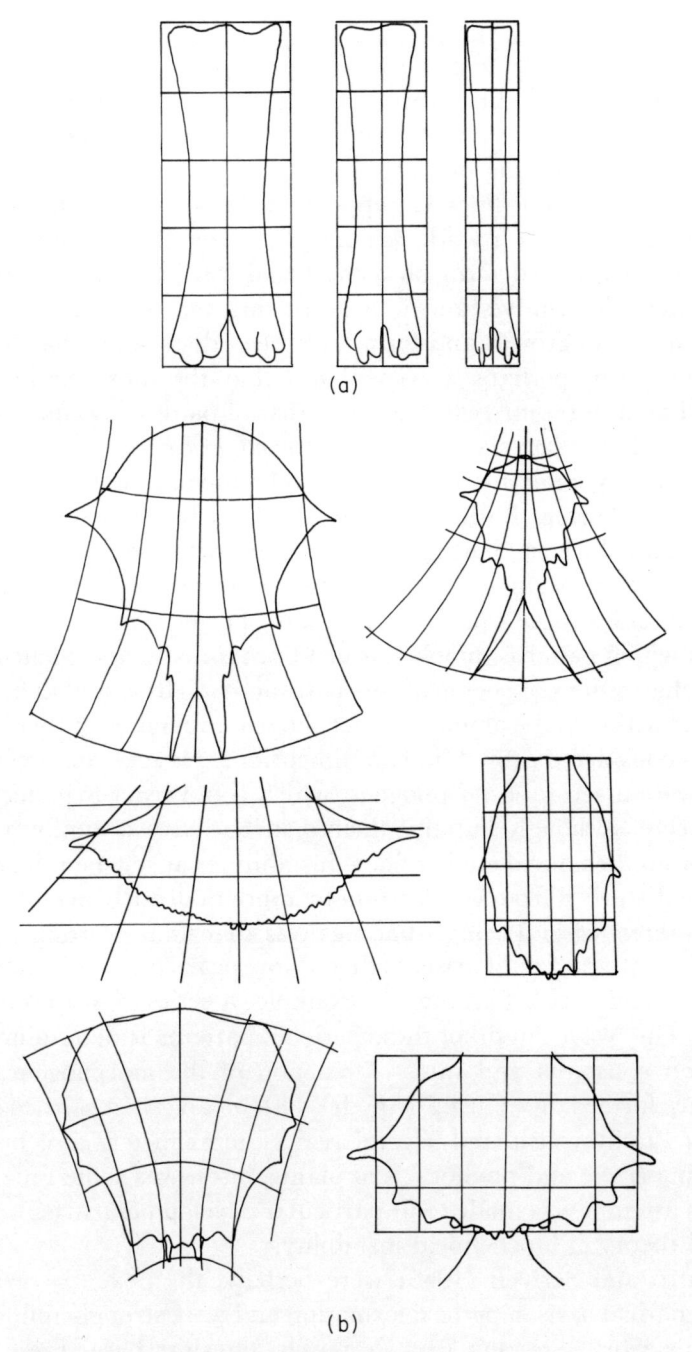

Figure 5.3 (a) Cannon-bones of (left to right) ox, sheep and giraffe. (b) Carapaces of various crab species. This variation in shape of basic patterns produced by using different coordinate systems was demonstrated by D'Arcy Thompson (1917).

sheep and ox is straightforward, whilst the transformations between the carapaces of various types of crab is more involved: 'a system of slightly curved and converging ordinates, with orthogonal and logarithmically interspaced abscissal lines . . . appears to satisfy the conditions'.

D'Arcy Thompson's work was unique, in that it attempted to put the study of growth and form on a mathematical basis. His studies were annoyingly descriptive, and it is surprising that so little work on the mathematics of growth and form have been done since this foundation was laid. It is perhaps a coincidence that the most mathematically studied area in recent years has been that of pattern specification, dealt with in only the last six pages of *On Growth and Form*.

The first dynamic mathematical developmental model was proposed in 1952 by Turing. Rather than simply describe particular patterns in mathematical terms, Turing was concerned with the actual mechanism in operation, envisaging the important parameters to be a system of chemical substances called morphogens. Turing was interested in the way in which patterns may be formed from initially homogeneous starting configurations of chemical concentrations. Cells would then differentiate according to the amount of morphogen impinging on their surfaces. Turing considered growth in two dimensions. He took an area in which there were a series of morphogens, which could react together and diffuse freely, although initially their distribution was uniform. Turing made a mathematical study of how instabilities arise when the system is slightly disturbed from equilibrium by random disturbances. One of the most interesting of Turing's findings was that under certain conditions periodic patterns could arise: high and low peaks of morphogen concentration could occur, forming, for example, a series of stripes of regular width. The 'wavelength' of these periodic patterns is determined by the diffusion constants and rates of reaction of the morphogens. Several patterns formed by Turing's model system are shown in Figure 5.4. Turing briefly considered several real systems in terms of his model, including *Hydra* and phyllotaxis in plants, but it was some time before a serious attempt was made to fit particular development structures to the overall theory of biochemical instability.

Gmitro and Scriven (1966) were perhaps the next to consider the mathematical basis of pattern formation and were strongly influenced by Turing's work. Starting from a purely physical base, these authors attempted to show that biological and physical patterns obey the same general principles. Their first models could therefore be considered as substitute physical system models and have already been discussed in the previous chapter.

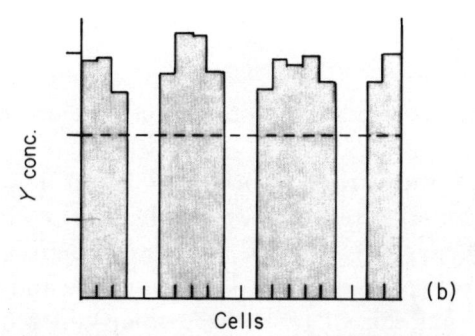

Figure 5.4 Patterns produced by Turing's model. (a) A two-dimensional dappled pattern. (b) A striped pattern in a ring of 20 cells. The pattern is represented by the concentration of a morphogen Y whose profile is indicated by the shaped area. The dotted line represents the initial starting concentration at which the level of Y is identical in all cells. The pattern is produced by a series of reaction rules involving morphogens X and Y together with a number of other chemicals which act as catalysts. The net effect of the reaction system is to transform X into Y whilst producing X and destroying Y at identical rates. Diffusion from cell to cell is interrupted when Y levels reach zero, thus stabilizing the pattern.

The basic theoretical system considered by Gmitro and Scriven consists of a membrane or thread of uniform thickness within which occur chemical reactions and diffusion. Exchange with the surrounding universe may also take place. At any location within the system each chemical species is said to obey the following 'conservation equation':

Rate of accumulation = net rate of + net rate of + net rate
within the system   production by   influx by   of input
                          chemical   diffusion   by exchange
                          reaction   along the   with
                          within the   system   surrounding
                          system   universe

In mathematical terms the equation becomes

$$\frac{C_i}{t} = R_i + J_i + Q_i$$

where C_i is the local concentration of the ion species and t represents time.

At a steady state, the various concentrations are by definition constant. If perturbation takes place, however, the rate at which the perturbation grows or decays is controlled by the competition between: (a) chemical reaction between system and surroundings and within the system, and (b) diffusion caused by concentration differences within the system. By using matrix notation and matrix multiplication (see below), it is possible to derive an equation which describes these factors, and the equation can be subsequently solved by the methods of harmonic analysis.

It had previously been shown by Fourier (see, for example, Bell, 1951) that a spatial pattern may be expressed as a weighted sum of members of a set of characteristic functions called *eigenfunctions*. It is at this point that the average biological reader may well throw up his hands in horror and despair at ever understanding how seemingly one-dimensional mathematical equations can ever represent spatial patterns. The answer lies in matrix theory.

A *matrix* is a rectangular array of elements that are members of a field or ring, and is made up of a number of rows and columns. For example,

$$A = \begin{pmatrix} a_{11} & a_{12} & a_{13} \\ a_{21} & a_{22} & a_{23} \\ a_{31} & a_{32} & a_{33} \end{pmatrix}$$

is a 3 × 3 matrix. If we have a series of biochemicals diffusing in a given space, then their interactions will be definable only in terms of matrix algebra. The reason for this is as follows. Take a series of biochemicals $X_1, \ldots, X_n$ diffusing in space. There is both interaction between the chemical species present and diffusion around the space. The concentration of any chemical at a given point in time and space will depend on the concentrations of all the other chemicals. For example, suppose that two biochemicals are being synthesized and destroyed at rates $A(X, Y)$ and $B(X, Y)$, each of which depends upon the concentration of X and Y at each point in the tissue, then the rate at which the biochemicals are diffusing throughout the tissue may be described thus:

$$\frac{X}{t} = A(X, Y) + D_1 \Delta^2 X$$

$$\frac{Y}{t} = B(X, Y) + D_2 \Delta^2 Y$$

where $D_1 \Delta^2$ and $D_2 \Delta^2$ are diffusion rates for X and Y respectively. This involves the solution of a set of partial differential equations for each chemical, which would result in a single equation of great complexity.

Matrix algebra has several advantages. Firstly, it allows a convenient shorthand representation of the system in mathematical terms, but more important, it allows the calculation of a set of parameters called *eigenvalues* which give the mathematician essential information about the way in which the system behaves. As described above, a spatial pattern can be expressed mathematically as a Fourier equation (the simplest of which describes a sine wave), the form of which is determined by a set of *eigenfunctions*. The particular eigenfunctions necessary to describe a given pattern are associated with particular sets of eigenvalues. There is thus a link between the biochemical conditions of a system (as abstracted by the mathematical modeller) and biological pattern. This link could be summarized in the following way:

Biochemical conditions
↓
Construction of matrix equation
↓
Calculation of eigenvalues
↓
Construction of Fourier equation to describe pattern

Gmitro and Scriven examined cases with one, two, and three biochemical species. In the first two cases it is possible to write an explicit equation for the eigenvalues. With three chemicals or more this becomes too involved. From the equation it is possible to find the relationships which the parameters of the system must satisfy in order for a given type of instability to occur. An *instability* is defined as a form of biochemical instability which gives rise to patterns of concentration of one or more of the chemicals in the system. It may be stationary or oscillatory. More types of instabilities are possible the more chemicals are present, and besides the biochemical conditions of the system, the shape of the system is also an important parameter: different patterns will be formed in a ring, sphere, or line.

SELF-ORGANIZATION IN NON-EQUILIBRIUM SYSTEMS

By constructing models of this general form, it is possible to show that small numbers of biochemicals could give rise to repeating patterns in

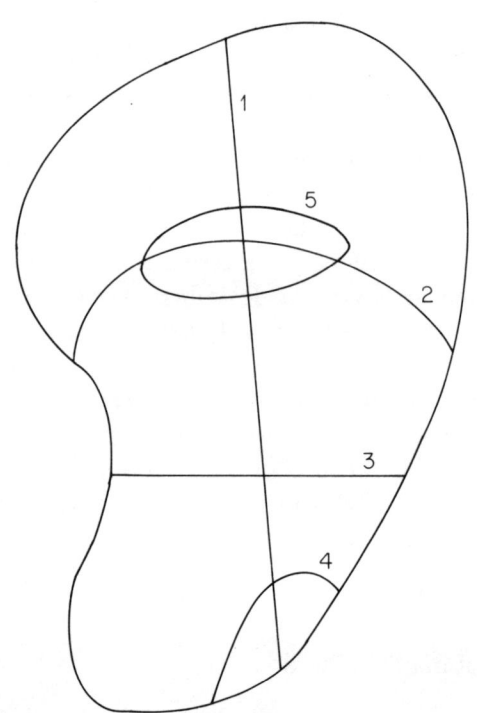

Figure 5.5 Compartment boundaries on the *Drosophila* wing imaginal disc numbered sequentially. After given stages in development, cells do not cross these lines. The disc is a basically two-dimensional structure.

two dimensions. Another approach to the mathematical description of general pattern formation has been used by Nicolis and Prigogine (1977), who again built on the foundation laid by Turing, namely analysis of the rate equations describing nonlinear interactions between at least two morphogenetic substances. Other authors have attempted to model particular development situations. Kauffman (1977) uses the overall framework of biochemical dynamical systems outlined above to explain the formation of compartment boundaries in *Drosophila* imaginal discs. The problem, briefly, is to explain how a two-dimensional plate of cells can be differentiated, as shown in Figure 5.5. Kauffman showed that the compartment boundaries fit well with the expected wave patterns formed by chemical systems generated on an ellipse. The geometries of the patterns are dictated by the shape of the tissue and a succession of different patterns may therefore be formed during development.

Bunow, Kernevez, Joly, and Thomas (1980) advanced Kauffman's mathematical analysis by using a computer to calculate the patterns that would be produced using particular input conditions. This numerical method showed that the pattern and sequence of the compartment boundaries could only approximate to the simulated lines produced, and that the approximation was worse rather than better when more anatomically accurate imaginal discs were simulated. Bunow raises one of the central criticisms of amending the model further:

'One might make a number of *ad hoc* modifications to the model which could improve its performance . . . Intuitively, it seems clear that an appropriate chosen inhomogeneity could produce any pattern whatever. Such a modification deprives the Kauffman model of its attractive simplicity, however, reducing it to an uninspiring exercise in curve fitting.'

Although the Kauffman model is certainly capable of generating the kinds of boundary found in imaginal discs, it is capable of generating almost *any* kind of pattern with a proper choice of parameters. There are other biochemical schemes which offer a more restricted interpretation, however, and one of the most well known of these was devised by Gierer and Meinhardt (1972). One of the major differences between the two approaches is that the Gierer and Meinhardt model is more specific regarding the interactions which may occur between the biochemical species present in the system.

Gierer and Meinhardt's basic mechanism involves two compounds interacting auto- and cross-catalytically on their own and on each other's production. One compound (A) activates and the other (I)

inhibits. Using such a mechanism, a variety of patterns can be formed, theoretically at least, by solving equations of the form

$$A = cA^2/I - \mu A + D_a \Delta A + C_0 \quad \text{(activator)}$$
$$I = cA^2 - vI + D_i \Delta I + C_1 \quad \text{(inhibitor)}$$

The change of A per time unit (A) is proportional to an autocatalytic term (A^2), is slowed with increasing inhibitor concentrations $(1/I)$, decays in a first-order reaction $(-A)$, and diffuses $(D_a \Delta A)$. C_0 represents a basal level of activator production. The change in I is a function of the cross-catalytic influence on A (A^2), of decay $(-vI)$ and of diffusion $(D_i \Delta I)$. C_1 represents a basal level of inhibitor production.

Simulations of this basic model are described in Chapter 8. They show that short-range autocatalytic activation in conjunction with larger-range inhibitory effects can lead to concentration patterns showing self-regulatory patterns like those seen in morphogenetic fields. The area of high activity is analogous to the classical 'organizer region' (see Chapter 3).

Meinhardt and Gierer have recently (1980) updated and refined their model to account for various forms of growth during development. One of the problems with their initial formulation was that it only accounted for the response of cells to the local concentration of a morphogen. What of the local interactions between neighbouring cells? A prime example of this sort of interaction occurs in so-called intercalary regeneration, where a tissue has an intermediate region excised (Figure 5.6). Cells at the cut edges of the tissue can often divide and 'fill in' the missing cells. Meinhardt and Gierer suggest that the following mechanism occurs: position can be specified by a physical 'positional value' r. A discontinuity in the pattern leads to a discontinuity in r. This leads to the production of a diffusible substance, which leads to a spatial average of r over a certain range. Signals responding to the discontinuities can be generated on this basis. Slack (1980) has put forward a discrete model of the same process. In this case, cells in the regenerating region contain a series of biochemical switches that can be either 'on' or 'off'. Switches in neighbouring cells may have different patterns of 'on' or 'off' states (Figure 5.7). The model proposes that in a tissue with a normal pattern of on and off switches there is no switch setting change, but if some cells are removed, diffusion may occur. Cells in the regenerating region have their switches reset according to the positions they take up.

Goodwin and Cohen (1969) suggested that positional information in developing systems can arise from the formation of a phase gradient.

The basic assumption of their model is that every cell in a developing field has the potential to oscillate with period T. As the oscillation progresses, the cell enters three states one after the other. Firstly, the brief emission of a repeating signal at (for example) times 0, T, $2T$, etc. Secondly, a long refractory period T_R occurs during which the cell is insensitive. Finally, there is a time interval $T - T_R$ during which the cell cannot emit signals, although it is sensitive to incoming signals. Goodwin and Cohen showed how an organizing wave could be set up by the reception and emission of signals in this way. If a gradient of oscillator frequencies is set up, a form of positional information arises. It is possible that under certain conditions the oscillator with the highest frequency can entrain the others—it would then correspond to the embryological concept of the 'organizer'.

After a signal S is released, a second, related pulse is produced. This

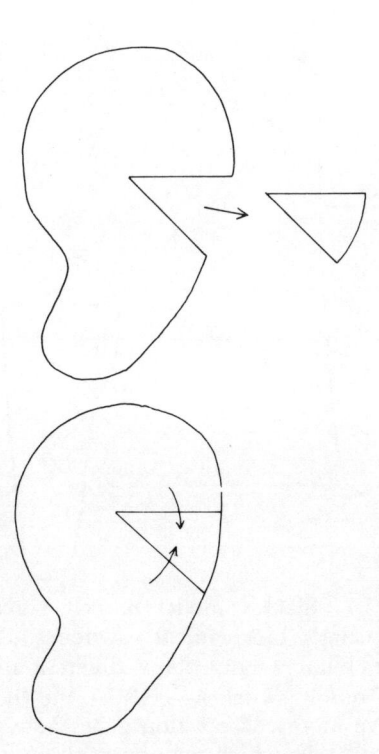

Figure 5.6 Intercalary regeneration in an imaginal disc. Removal of a segment leads to production of the missing part by neighbouring cells.

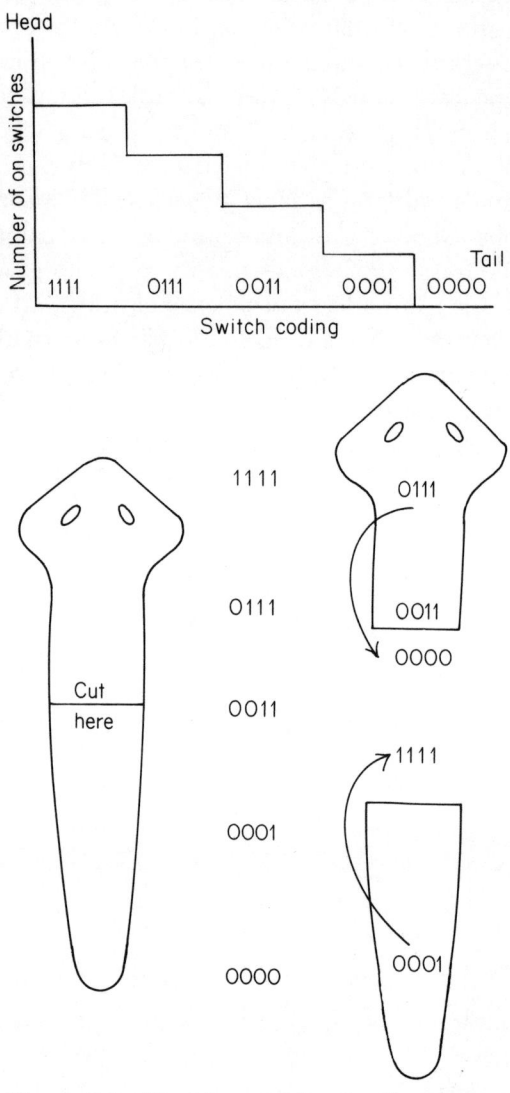

Figure 5.7 Slack's model of cell regeneration using biochemical switches. The possible binary codes show different numbers of on/off switches—cells at the 'head' end have more cell position genes activated than 'tail' cells which may have all genes in the 'off' position. When regeneration takes place, neoblast cells have their code reset when they migrate to the cut surface.

propagates through the field, establishing a phase gradient. Although experimental evidence of the general working of this kind of model in real developing systems is lacking, the aggregation of cellular slime moulds shows many of the properties of oscillatory systems (Darmon, Brochet and da Silva, 1975). For example, mutants of the slime mould *Dictyostelium discoideum* that do not aggregate normally can be made to do so by subjecting the individual amoebae to periodic pulses of cyclic AMP (cAMP). Conversely, continuous flow of cAMP fails to promote differentiation.

CATASTROPHE THEORY AND DEVELOPMENT

A succinct definition of catastrophe theory has been given by Zeeman (1976):

'Catastrophe theory is a new mathematical method for describing the evolution of forms in nature. It was created by Rene Thom who wrote a revolutionary book *Structural Stability and Morphogenesis* in 1972 expanding the philosophy behind the ideas. It is particularly applicable when gradually changing forces produce sudden effects. We often call such effects catastrophes, because our inhibition about the underlying continuity of the forces makes the very discontinuity of the effects so unexpected, and this has given rise to the name. The theory depends upon some new and deep theorems in the geometry of many dimensions, which classify the way that discontinuities can occur in terms of a few archetypal forms: Thom calls these forms the *elementary catastrophes*. The remarkable thing about the result is that, although the proofs are sophisticated, the elementary catastrophes themselves are both surprising and relatively easy to understand, and can be profitably used by scientists who are not expert mathematicians.'

As Saunders (1980) states, catastrophe theory is especially useful when describing large and complex systems (in a similar way to the use of matrices with the corresponding eigenvalues as described earlier). Catastrophe theory can allow us to predict much of the qualitative behaviour of the system without even knowing what the differential equations describing the systems are, much less solving them.

What is a catastrophe? There are a series of seven elementary catastrophes which can be represented by a series of shapes in space. They are the *fold, cusp, swallowtail, butterfly, elliptic umbilic, hyperbolic umbilic,* and *parabolic umbilic*. There are a number of simple examples which have been used to illustrate the various types of catastrophe in action. These are often so simple as to hinder rather than help the non-mathematician's comprehension of catastrophe theory. The general concept of a catastrophe can be inferred, however, by making a 'catastrophe machine' as described in Appendix 4.

One of the principal theorems of importance to developmental biology arising from catastrophe theory is due to Zeeman (1974). This is that when differentiation of a mass of cells occurs into two types, the boundary between the two cell types always forms to one side of its final position and then moves through the tissue before stabilizing in its final position. The starting point for this hypothesis is the definition of four important parameters in the development of a tissue E during time T, namely:

1. *Homeostasis*. Cells are in an equilibrium which may change with time.
2. *Continuity*. The chemical, physical and dynamical conditions in different cells may be represented by smooth functions on E.
3. *Differentiation*. At the beginning of T there is only one type of cell, at the end there are two, with no continuous variation one to the other.
4. *Repeatability*. Development is stable—a similar development would take place under small perturbations of the initial conditions.

From these parameters, Zeeman proposes the following theorem. *Homeostasis, continuity, differentiation and repeatability imply the existence of a primary wave*. In other words, *a frontier forms, moves and deepens, then slows up, stabilizes, and finally deepens further*. This theorem is of course qualitative, and gives no indication of whether or not the wave is visible or hidden.

To examine the proof of this theorem requires a deeper knowledge of the theory than the average biologist will possess, and the mathematically inclined reader is referred back to Zeeman (1974). It *is* possible to show at least some of the reasoning behind the theorem in non-mathematical terms, however. Consider Figure 5.8(a). This shows a two-dimensional space representing a line S placed orthogonal to the hypothetical frontier, together with a time axis T. At a certain time a discontinuity will appear to separate the two cell types. This is shown at the right-hand side of the diagram. A cell α on the boundary (shown arrowed) will be undifferentiated through time until it changes state. Can catastrophe theory explain what happens? The first step is to fill in the missing part of Figure 5.8(a). The simplest type of catastrophe is the fold, but this does not fit, so the next most complex form, the cusp, is tried. A simple cusp is shown in Figure 5.8(b). This will not do, how-

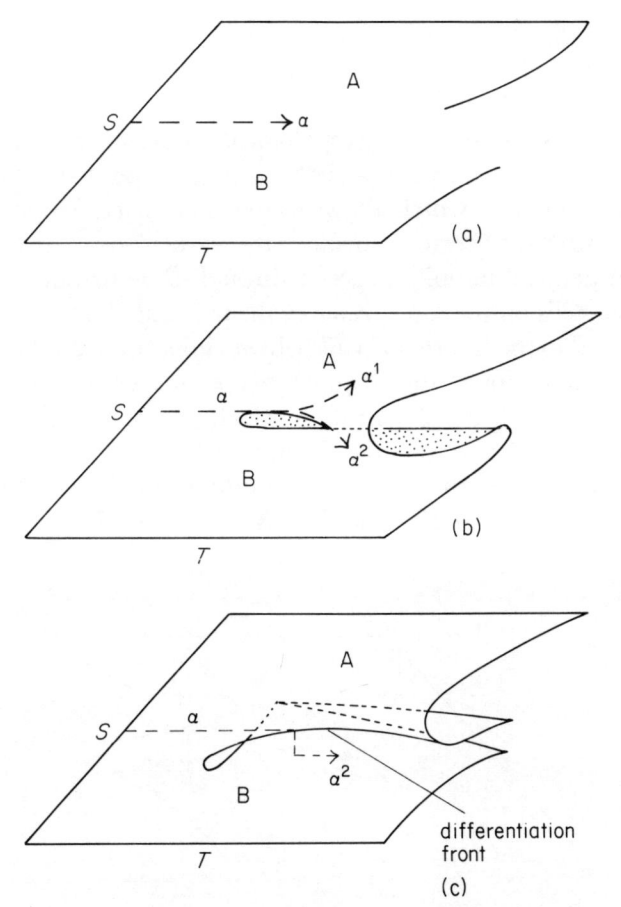

Figure 5.8 An explanation of differentiation patterns using catastrophe theory (after Zeeman, 1974). (a) A cell α progressing through time T is to change state. (b) A simple cusp catastrophe may govern differentiation into state α^1 (tissue A) or state α^2 (tissue B). (c) The more complex cusp here curves back on itself. This has the effect of making it structurally stable and also stops the frontier between A and B moving right out of the tissue.

ever, as we want points close together to have the same differentiation—in this case there is an 'all or nothing' differentiation where cells close to the cusp can be either A or B cells. Cell α then has a direct choice between α^1 (A type) or α^2 (B type). The cusp must therefore curve so as not to be parallel to the T-axis. The cusp will also have to curve backwards to stabilize the frontier and prevent it moving on to the far end of the tissue (see below). The final pattern therefore requires amendment and appears as in Figure 5.8(c).

It is now possible to use this topological model to trace the history of cells in the tissue. We can do this by using a modified form of Figure 5.8(c) (see Figure 5.9) which allows us to look at what happens with (top surface) or without (bottom surface) the catastrophe. Any cell on the bottom surface will happily proceed through time without any change. In addition, cells on the top surface in the intervals X and Y will also not change. A cell in the interval Z will behave differently. No differentiation will occur until it crosses the cusp. At this point there is a sudden change in state and the cell develops into a B cell. No state change will have occurred in the tissue until time T_1, at which time discontinuity occurs until a stable concentration pattern is produced at time T_2. Note that the position of the discontinuity also moves from β_1 to β_2.

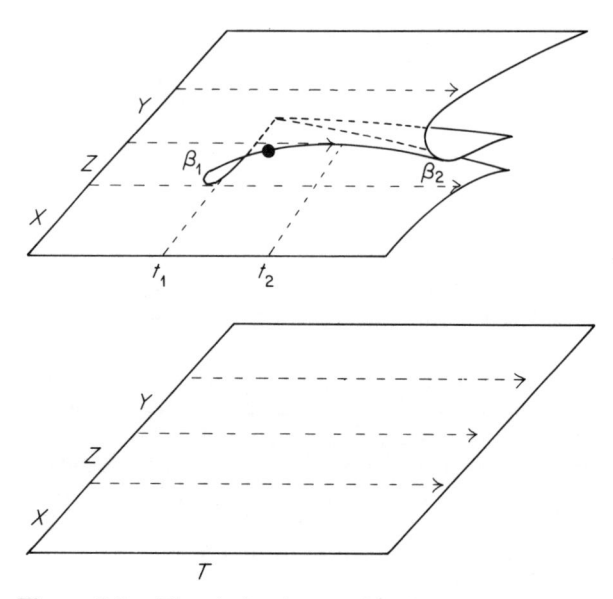

Figure 5.9 The behaviour of cells in different regions of the tissue. Cells in region X or Y will differentiate into B or A type cells respectively at time T_1. Cells in region Z form B cells from time T_1 onwards, but the more distal cells (closest to Y) become differentiated at a later time. Thus the frontier appears to 'move through' the tissue.

The reader will probably need to examine Figures 5.8–5.9 carefully to convince himself of this result. In the theoretical form it does not seem particularly useful. The test of the model is in its practical verification and this has taken several forms. Cooke and Zeeman (1976) attempted to explain the formation of the repeated pattern of vertebrate somites by means of a primary waveform like that described above that interacts with an oscillator in each cell. There are of course problems. We cannot 'see' cell states, only the differentiation which accompanies physical differences in cells. It is therefore difficult to show the validity of Zeeman's theorem directly. The only evidence so far is rather circumstantial data from heat-shock perturbations of amphibian embryos (Elsdale, Pearson, and Whitehead, 1976). These authors suggest that heat shocks perturb the oscillator; anterior–posterior asymmetries induced in somite segmentation might reflect the gradual restoration of the original oscillations following a disturbance. It is further noted that the lag between heat shock and visible effect suggest that the wavefront–oscillator interaction takes place well before segmentation: this seems consistent with the Cooke–Zeeman model.

What, then, in retrospect can catastrophe theory offer to development biology? In contrast to the non-equilibrium systems of Nicolis and Prigogine, catastrophe theory allows dynamics and form to be intimately related. This is certainly of interest to the mathematican, but its applications do seem rather vague. It is difficult to see how anything other than generalizations can come out of the theory, as it is practically impossible to feed real parameters into the topological models.

CONCLUSIONS

This chapter has taken us from the arithmetical forays of D'Arcy Thompson to the almost alien complexities of catastrophe theory and dissipative structures. It had been impossible in a book written for non-mathematicians to do more than hint at the methods and uses of such complex techniques. I hope that sufficient information has been given to leave the reader with some kind of feel for the gamut of mathematical models in developmental biology.

Are there any lessons to be learned from mathematical models in general? It is confusing for the non-mathematical biologist to be confronted with two or three branches of mathematics which purport to model the same process, and certainly there seems to be scope for greater dialogue between mathematicians in the various fields to decide amongst themselves what sort of mathematics should be used to model what kind of process. The biologist is left with the feeling that the mathematician deigns to leave his ivory tower for a little while to lower himself to frivolous *applied* matters, only to retreat at the first sign of disproof of his model.

In the same way that biologists become entrenched in their own research field, mathematicians must also become limited in their ability to see problems in anything but the terms of their own pet mathematical techniques. The future of mathematical models in developmental biology will depend on some literate and able entrepreneurs homing in on the important features of development and explaining them in the correct mathematical framework. Whether differential equations or catastrophe theory are the correct methodologies is an open question which will perhaps be resolved before this book is out of print.

REFERENCES

The following books and articles contain general reviews and descriptions of the particular mathematical approaches described in this chapter.

Gierer, A., and Meinhardt, H. (1972) A theory of biological pattern formation, *Kybernetics*, **12**, 30–39.

Gmitro, J. I., and Scriven, L. E. (1966) A physicochemical basis for pattern and rhythm, in: *Intracellular Transport*, Academic Press, N.Y.

Goodwin, B., and Cohen, M. H. (1969) A phase shift model for the spatial and temporal organization of developing systems, *Journal of Theoretical Biology*, **25**, 49–107.

Nicolis, G., and Prigogine, I. (1977) *Self Organization in Non-equilibrium systems*, Wiley–Interscience, N.Y.

Saunders, P. T. (1980) *An Introduction to Catastrophe Theory*, Cambridge University Press, Cambridge, UK.

Thom, R. (1975) *Structural Stability and Morphogenesis*, Benjamin, N.Y. (translation by D. Fowler from the French).

Thompson, D'Arcy W. (1917) *On Growth and Form*, Cambridge University Press, Cambridge, UK.

Turing, A. (1952) The chemical basis of morphogenesis, *Phil. Trans. Roy. Soc. B*, **237**, 37–72.

Other papers mentioned in the text may also provide useful further reading, especially the papers by Mitchinson, Cooke and Zeeman, and Slack.

Bell, E. T. (1951) *Mathematics—Queen and Servant of Science*, McGraw-Hill, N.Y.

Bunow, B., Kernevez, J.-P., Joly, G., and Thomas, D. (1980) Pattern formation by reaction–diffusion instabilities: application to morphogenesis in *Drosophila*, *Journal of Theoretical Biology*, **84**, 629–649.

Cooke, J., and Zeeman, E. C. (1976) A clock and wavefront model for control of the number of repeated structures during animal morphogenesis, *Journal of Theoretical Biology*, **58**, 455–476.

Darmon, M., Brochet, P., and da Silva, L. H. (1975) *Proc. Natl. Acad. Sci. (USA)*, **72**, 3163–3165.

Elsdale, T., Pearson, M., and Whitehead, M. (1976) Abnormalities in somite segmentation induced by heat shocks to *Xenopus* embryos, *J. Emb. Exp. Morphol.*, **35**, 625–35.

Kauffman, S. A. (1977) Chemical patterns, compartments, and a binary epigenetic code in *Drosophila*, *American Zoologist*, **17**, 631–648.

Meinhardt, H., and Gierer, A. (1980) Generation and regeneration of structures during morphogenesis, *Journal of Theoretical Biology*, **85**, 429–50.

Mitchinson, G. (1977) Phyllotaxis and the Fibonacci series, *Science*, **196**, 270–275.

Slack, J. (1980) A serial threshold model of regeneration, *Journal of Theoretical Biology*, **82**, 105–140.

Zeeman, E. C. (1974) Primary and secondary waves in developmental biology, *Lectures on Mathematics in the Life Sciences*, **7**, 69–161, American Mathematical Society, Rhode Island.

Zeeman, E. C. (1976) Catastrophe theory, *Scientific American*, **234**, 65–83.

Chapter 6

Computer modelling and development

Although modelling in general terms seems to be a 'respectable' pastime for embryologists, there is a gulf between paper-and-pencil and substitute system models on the one hand, and mathematical and computer models on the other. Many biologists are wary both of theorizing and of their theoretically orientated colleagues. There are good reasons for this. Whereas a practical biologist is limited by the rules of logic and deduction regarding his experimental results, the theoretician has few such worries, inhabiting as he does the never-never land between Art and Science. If he wishes to build a model of kidney development using invisible gnomes who carefully place each cell in position, then who is to argue?

There is no doubt that a number of developmental computer simulations are not far removed from this ridiculous example, but I hope to show in this and the following three chapters that a high proportion of computer models are worth both further study and extension. Our first task is to expand on the brief description of a computer model given in Chapter 2. The computer-wise reader should therefore turn to page 71.

COMPUTERS AND MODELS

For the purpose of this book, the word *computer* should be taken to mean *digital computer* unless otherwise stated. (The other major type of computer, the *analog computer*, has not been greatly used in simulations of developmental processes.) A digital computer produces a solution to a particular problem only if it is presented with a series of instructions that it is able to perform. These instructions must ultimately be written in the machine code that can directly access the computer's memory stores, processing unit, and input/output devices. For most purposes, a *high-level computer language* is used for writing the instructions or *program*, and the computer has a special *compiler* that can translate the high-level language into machine code.

Typical high-level languages are ALGOL, FORTRAN, and BASIC.

Any computer operation is really a simulation, even a simple addition such as 1 + 1 = 2 (the computer is simulating a paper-and-pencil process). A slightly more complex computer program to add, say, three numbers together, might look as follows:

```
BEGIN
INTEGER X, Y, X, A
READ (X)
READ (Y)
READ (Z)
A = X + Y + Z
PRINT (A)
END
```

The essential elements of the program are therefore instructions to start, stop, read in data, print out data, and to do manipulations within the computer.

A computer is really a gigantic array of pigeon holes. Each can be empty or full, and a line of such 'two-state' holes or locations can be made to represent a number in binary notation (see page 98). The instruction INTEGER X, Y, Z, A in the above program tells the computer that four numbers will be used in the program, and will therefore need 'holes'. The computer therefore assigns four sets of locations for the numbers, and 'tags' each so it will be able to find them on recall. The numbers X, Y, and Z are called variables and are read in and stored. The instruction A = X + Y + Z tells the computer to add the contents of locations X, Y, and Z and to store the sum in location A. If the program had been written so that A was printed *before* the A = X + Y + Z instruction, the value 0 would have been obtained—A would be empty at that point as programs run sequentially.

Computers can do more than simply add. Their ability to move the contents of locations around makes them especially useful for developmental simulations. A particular advantage of this facility is that locations can be *spatially referenced* to one another. Instead of calling input variables X, Y, and Z, they may be termed X(1), X(2), X(3). There is little advantage in reading just three numbers into an *array*, as the indexed series is called. Imagine, however, a more complex version of the program example where 100 numbers are to be inputted and added together. Rather than specify 100 separate locations, the computer can set up an array with 100 locations as follows:

```
BEGIN
INTEGER ARRAY X(1:100)
```

```
INTEGER A, B
A = 0
B = 0
B = B + 1
READ X(B)
A = A + X(B)
IF B LT 100 GO TO 1
PRINT (A)
END
```

This second program uses several more sophisticated elements than the first example. Note that the size of the array is initially specified so that the computer can label the appropriate number of locations. There is also a jump instruction 'If B is less than 100, go to 1'. This has the subsequent effect of reading and entering the next array location element and entering it into A. When all array values are entered and added, A may be printed out as the sum.

The value of the array system, which can be one-dimensional as in this example, or two-, three- or even multidimensional, will be further discussed in the next chapter.

THE MODEL SIMULATION

The design of a model to be run on a computer will take several stages. Firstly, the problem must be defined in a series of clear statements. It must also be understood what the model is expected to achieve—it is little use preparing a large simulation with reams of computer output if the output tells nothing about the workings of the real system. (The process of abstracting the most important factors of the system to use in the model is an art in itself, and often requires a great deal of practice.) Secondly, the model is constructed from the abstracted statement. This stage involves making a flow chart of some kind, although the chart need not be as formal as a layout of the complete computer program with every individual instruction written in.

A flow chart is useful for several reasons. It enables communication of a computer program between users, or between user and programmer, if the running of the program requires some specialist help. (It is difficult to understand someone elses raw program, especially when written in an unfamiliar computer language.) It also enables the programmer to see how much time is spent by the computer in processing various parts of the program. It may even be possible to be given an estimate of the time a model will take to program. Although it is not easy to work out how long a program will take to develop and run, the number of branching instructions in the program give an idea of the 'logical complexity' involved. If a standard amount of programming time is associated with each branch instruction, based on past experience, an estimate of the total programming time can be obtained (Smith, 1968).

An important point to make at this juncture is that the mere formal construction of a computer model, before any computing is done, can give insight into the workings of the real system, and may even reduce or obviate the need to actually run the model on the computer. This is all to the good—few models are ends in themselves, and whether or not a full-scale computer model needs to be actually run on the computer really depends on the ability of the simulation to give meaningful information about the system being studied.

After the flow chart has been completed, the preliminary programming is begun. If the cost of computer time is no object, as is the case in many non-commerical computer installations, the modeller may be tempted to try to run the program before completely understanding what it is supposed to do (and, indeed, what it is *capable* of doing). Before beginning any detailed programming it is important to know the capabilities of the machine to be used. Although cost may be no object, time may be, and computers vary significantly in their processing speeds and turn-round times. There are also additional considerations to be taken into account, such as the advantages of using multi-access systems, or specialized input and output facilities. Multi-access systems allow interaction between user and program whilst the program is running, and this can be very useful in some simulation work. Specialized modes of input and output are considered in greater detail in Appendix 1.

The computer language used really depends on the requirements of the individual simulation. I have briefly mentioned the separation of computer languages into simulation and standard high-level languages. The main argument against using an ordinary computer language for simulation is that a computer processes instructions one by one—in a normal high-level language this is quite literal. A simulation language may itself be able to process instructions and handle data in 'pseudo-parallel', and so give a better approximation to real processes. Cells in real tissues do not divide 'one after the other'!

DETERMINATION AND RANDOMNESS

We have considered what a computer program looks like, and the important points to remember before embarking on the construction of a computer model. It is equally important to decide on the features of developmental systems that have to be considered when deciding on the parameters that are to be built into the model. One of the most important factors is the relationship between determination and randomness in developing systems.

We can separate all processes into these two categories. A *deterministic* process is one which is designed to operate in a previously determined manner. The overall pattern of normal development could be termed a deterministic process: it starts with the fertilized egg, and finishes with the production of an adult organism. A frog egg makes a frog, not an elephant. On the other hand, some processes are *random*. An example is the choice of sex in many organisms: there is an approximately equal probability that an X- or Y-bearing sperm may fertilize the egg, giving a male or female. In making a model of a given system it is therefore important to decide which elements are deterministic and which are random. In some cases this is not an easy decision to make, as examples throughout this book will show. There is, however, a partial solution to the dilemma.

As described by Tocher (1963), a parallel to the sort of complex situation we are confronted with in analysing developmental systems was reached by the mathematicians von Neumann and Ulam, when faced with the problem of trying to find solutions for the diffusion equations in physics. In diffusing systems, particles behave in a partly regular, partly irregular manner—by using statistical approximations to average over the particles, the random element was eliminated, and a deterministic description could be given.

This elimination of randomness in helping us to describe complex processes, where we could not hope to use a direct deterministic method, is often called the *Monte Carlo* or *random walk* technique. The making of random choices fitting overall statistical patterns is also well suited to simulations of large cell populations in biology. The majority of the computer models to be described in later chapters will be seen to use this methodology. An essential element in such simulations is the use of computers. As randomness is such an important factor, several repeat runs of a Monte Carlo simulation have to be done in order to get a measure of the repeatability of a result. (The simulation to be worked through in detail in Chapter 9 will be found to illustrate this point.) If the model is of any size at all, it will probably be too time-consuming to entertain doing even one run by 'hand'. The speed of the computer allows many thousands of operations to be carried out per second, making it ideal for Monte Carlo-type simulations.

The use of randomness to represent 'deterministic behaviour' can be illustrated by the following example. A cell may divide in a particular direction because there is less physical resistance from the surrounding medium for this division direction. Now it may be that there will be 'random' fluctuations in physical resistance throughout the medium. In any one cell division, the division directions will not be random, but will be caused by the difference in resistance. The cell division patterns overall will appear random, however, so it does not matter if one of the cells divides to the left, or to the right, so long as all the cells make the choice of division direction randomly. In other words, the use of randomness does not mean that the process being modelled is actually random *in vivo*, but that we cannot in practice work out the deterministic behaviour of the real system. This is a more practical statement of the 'levels of organization' doctrine quoted on page 9. At the single-cell level, the determinism of the cell division directions is important, but when the whole population is considered, it is the overall behaviour patterns produced by the cells acting in concert that is most interesting.

The random choice can be made by deciding on the number of alternatives and by using a *look-up table* to determine the outcome. If a certain random number is picked, then the look-up table decides the outcome, which part of the computer program is to be processed next. The flow chart in Figure 6.1 shows how this can be accomplished.

As the results obtained from a Monte Carlo simulation are necessarily dependent on the sequence of random numbers generated, it is important to have a reliable source of random numbers. The computer itself can generate 'pseudo-random numbers'. (Pseudo-random numbers are generated in a cycle, but as the cycle length is of the order of millions, it should not affect the simulation.) This method is suitable for all the simulations to be discussed here. Most computer installations have 'house' software for producing pseudo-random numbers, but Tocher (1963) describes methodologies. A generator of pseudo-random numbers should produce numbers evenly distributed over the whole range, and with no correlation between adjacent terms.

Another component of this type of simulation is the computer list structure (basically a modified one-dimensional array). When modelling large numbers of units, like cells, it is often important to maintain and update a lot of information, whether it is about individual units or the whole population. We will see later how this is especially useful when maintaining lists of which cells have divided in a particular 'generation'. This sort of operation can be done using a high-level language like **ALGOL** or **FORTRAN**. The universality of these languages is of great advantage, and they can often be used with success, although specialized

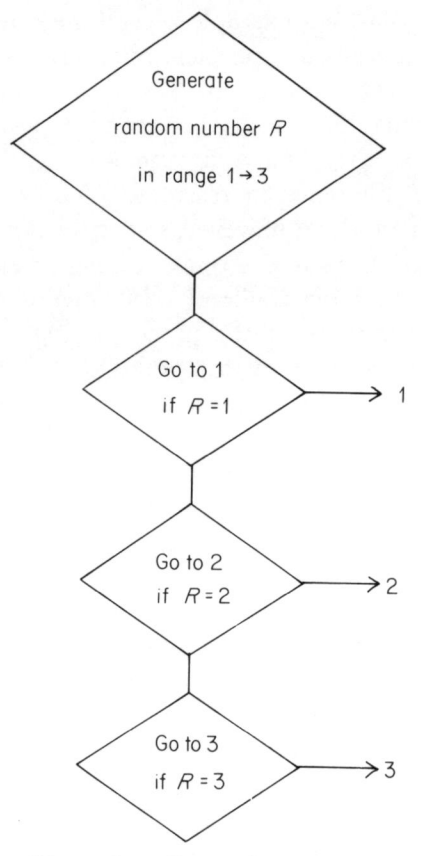

Figure 6.1 Part of a computer program flow chart which acts as a 'look-up' table. It processes the various choices of outcome after a random event.

simulation languages may sometimes be necessary. Most of the simulations described here have been carried out using ALGOL-based languages.

THE EXPERIMENTAL APPROACH

One of the bases of modern biology is the experimental approach, and developmental biology is no exception. Experiments on developing systems may take many different forms. We can look at mutants which develop abnormally, we can remove cells from one part of the embryo to see if the remaining section is still capable of normal growth, we can upset normal growth by adding growth inhibitors. The accent on all these approaches is that of *perturbing* a normal system to try to find out more about how it works.

In a similar manner, computer models can be easily perturbed by changing parameters, or by rewriting sections of the computer program. In this way it may be possible to see if these kinds of manipulations in computer models can be made to mimic the developmental systems which the models are designed to simulate. The logical basis of perturbations in real systems may therefore be analysable in terms of programmer-induced perturbations in simulations of the system.

Let us look at an example of this kind of approach in the analysis of a mutation with a well-defined effect—the mutant *bicaudal* in the fruitfly.

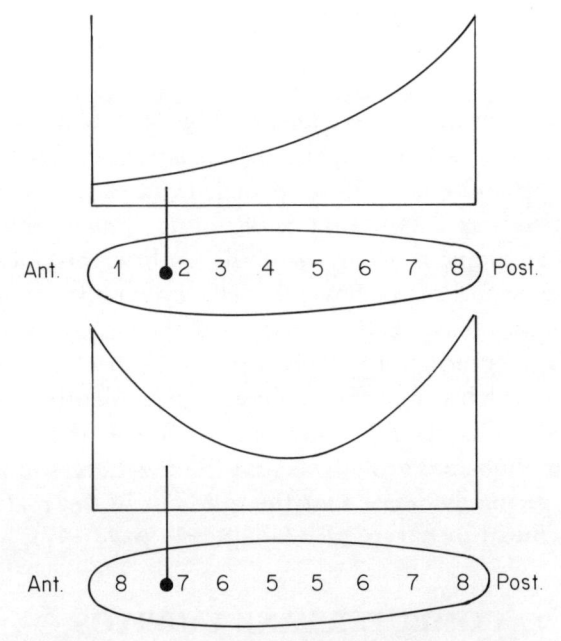

Figure 6.2 The experimental approach to model construction. A gradient of some character is proposed with a 'high point' at the posterior end. A cell marked as a solid circle in the normal embryo would form '2' structures. The *Drosophila* mutant *bicaudal* has two posterior ends, and the gradient profile would be as shown in the lower diagram. The cell marked as a solid circle would form '7' structures in this case.

Flies carrying the mutation may produce embryos with two posterior ends—such embryos are palindromic (and of course cannot develop into adults). The paper-and-pencil model proposed by biologists to explain this mutation is that there is a gradient of position in the normal embryo with a high point (source) at the posterior end and low point anteriorly (see Figure 6.2). Cells would then develop according to their position along the gradient, which they would interpret in some manner.

If the *bicaudal* embryo, instead of having a single posterior source has two sources of the position gradient, one at each end, the interpretation of the gradient by the cells will give a palindromic embryo. This much is quite straightforward, and to build a simple model of this situation hardly requires a computer. But what if we want to go further and try to find out what the gradient consists of? A computer simulation might be set up based on diffusion of a biochemical substance. Investigation of the properties of a gradient of this kind might allow us to show whether or not the diffusible substance could provide a suitable gradient mechanism.

Actual simulations along these lines will be described in Chapter 8 but the basic method is rather simple: all that is required is extra data on how the real system behaves if perturbed. Data of this type are readily available. Experiments show that constricting embryos by tying a fine hair somewhere along the length of the embryo produces abnormal growth. These experiments have the effect of preventing passage of biochemical signals between the two parts of the embryo on either side of the ligature. If the gradient model is correct, the gradient would be split into two halves which would then behave independently. What happens if we set up a 'diffusible substance' simulation in the computer? The results fit with abnormal growth seen in the two halves of the ligatured embryo, thus giving evidence that the gradient in the real system may well be biochemical in nature (see Chapter 8, page 147).

COMPUTERS AND EMBRYOS

The *bicaudal* example may well seem to be a sophisticated way of linking the computer to the testing of biological theories, but it errs very much on the side of simplicity. Biochemical gradients are conceptually easy to understand, but what about situations where three-dimensional masses of cells are dividing and differentiating into structures as complex as a human arm or leg? Can the computer solve problems that the human brain cannot? A naive answer to this question is 'no', because we all know that a computer can only follow the instructions fed into it. But is this really the case? In Chapter 8 a series of abstract models will be described, among them a series of patterns generated by the mathematician Stanislav Ulam. Ulam's patterns concern the growth of figures in a two-dimensional matrix, and growth of abstract 'squares' on the matrix occurs by following a series of very simple rules. A typical pattern and the instructions for its generation are shown in Figure 6.3.

(a)

(b)

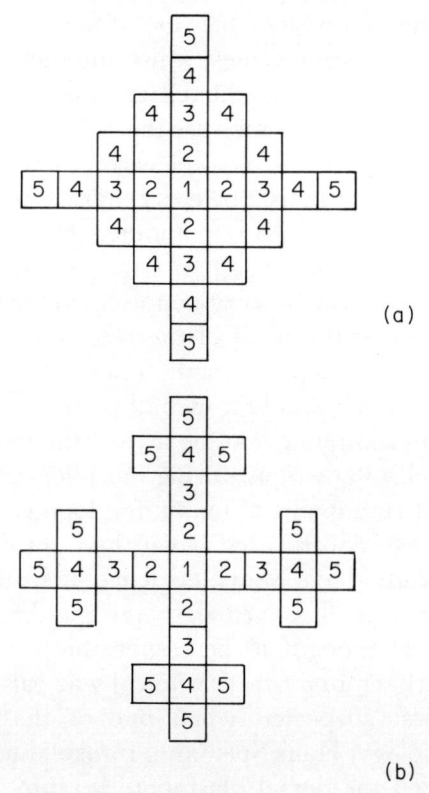

Figure 6.3 Different patterns grown by successive, replications of a single 'square' (marked 1) in two dimensions. The square could be imagined to be on a chessboard, and at each time step all non-diagonal squares adjacent to a full square become filled, unless: (a) they would touch more than one other filled square; (b) they would touch the *corner* of more than one filled square.

By feeding simple rules into the computer, a complex pattern emerges, suggesting that the computer may turn up some surprising results. It would certainly not be possible intuitively to see the end result from the initial rules unless one worked through the growh of the model, and out of the two alternative methods—computer or pencil-and-paper—the computer is much faster and less biased towards the preconceptions of the experimenter. Computer models can therefore be designed to do things that no scientist would attempt by hand, and they are especially useful in developmental biology because of the massive number of components, be they cells or molecules, which interact to give even a simple type of growth. Chapter 3 considered the building blocks of development, the most important of which are the cells, so it seems reasonable to assume that, in the same way as a television picture is constructed of a series of lines, a living organism owes its characteristic shape to the way in which its cells interact with one another. How can we analyse how these millions of cells work together? Many scientists believe that the rules of development will be very simple, and that doing this or that experiment will reveal the basis of the rules underlying development. But what if these rules depend on the interaction of large numbers of cells? It cannot be easy to hold a mental picture of cell interactions on this scale, and the computer may be a valuable tool in unravelling the complexities of cell interactions during morphogenesis.

The traditional standpoint of many developmental biologists is that the theory will largely work itself out if the right experiments are performed. We already have some evidence that this is a dangerous attitude. In the last fifty years there has been a search for a particular chemical thought to be responsible for the specification of pattern in the early embryo: this chemical was called the 'inducer'. Its existence was first suspected when one of the fathers of modern developmental biology, Hans Spemann, transplanted part of an amphibian embryo called the dorsal blastopore lip into a different part of a second embryo. The first embryo died, but the second embryo developed with two heads and nerve cords (see Figure 6.4). It seemed simplest to explain this phenomenon in terms of a chemical called the inducer, which *induced* the formation of the second axis of symmetry in the host embryo.

Unfortunately for developmental biology, it was found that a whole series of chemicals would do the same job as the inducer, and much of the present century's research into embryology has involved adding more and more esoteric chemicals to embryos to see what bizarre developmental patterns can be produced. The problem with research of this kind is

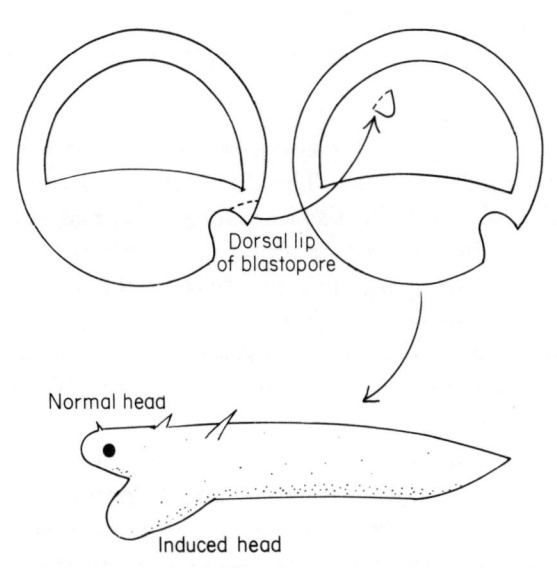

Figure 6.4 Induction of a secondary embryo by use of a Spemann graft.

that things may not be as simple as the original workers construed. Even with our present ability to detect tiny concentrations of biochemical fluctuations, we are no nearer to finding the true nature of the inducer than was the last generation of embryologists with their relatively crude assay techniques.

Let us look at this problem from a theoretical point of view. A particular stimulus applied to a system produces a particular effect. To start with, the stimulus could act in one of two ways: it could merely be a trigger setting off reactions in the host tissue (in which case identification of the stimulus would only be of limited interest), or it could be capable of directing the neighbouring cells to form the 'ground plan' of the secondary embryo. This is what the early embryologists may have been hoping in their great search for the inducer. The 'trigger' hypothesis has been largely ignored, and one suggestion for its lack of favour (at least in the past) is that a mechanism of this kind must be rather more complex that the 'ground plan' theory. Think back to the 'biochemical gradient' model suggested for the *bicaudal* mutation. If the inducer acts in the way that was suggested for the insect embryo, providing a 'ground plan', it can be simply explained: different levels of the inducer programme cells in different ways, so that the full range of structures in the secondary embryo are formed. If the inducer merely triggers the cells to respond, then a different kind of mechanism must be acting.

WHAT CAN AND WHAT CAN'T BE SIMULATED?

This chapter has suggested that heuristic models are most amenable to computer study, and most of the computer models discussed in the later chapters fall into this category. The applications of computers, within the confines of this book, have also been largely limited to those heuristic models involving the interaction of large numbers of 'simple units' (cells). For reasons of space and the bias of the author, subcellular processes have been neglected, when there is actually no doubt that they play a real and important part in development. After all, some whole organisms are unicellular, and therefore have no cell–cell interactions (see Chapter 3).

We will look at many computer models of developing systems in the next two chapters, and some of them have been found more amenable to computer analysis than others. There is no hard-and-fast rule as to what can, and what can't be simulated, but some guidelines might be set out in the form of questions:

1. Am I trying to simulate a process that takes place in more dimensions than I have available (for example, modelling growth of a three-dimensional limb bud in two dimensions, or modelling diffusion over an area and using the equations for diffusion in a line)?
2. I have a complex organism. Is there a simpler living system which exhibits the property I'm trying to model? Would it be easier to consider this system than to try and make a more complicated model?
3. Is the model likely to give us insight into the workings of the real system? (It may be impossible to tell this straight away, but a little soul-searching should make sure that no obviously trivial model is run.)
4. How many speculative assumptions do I have to make in setting up the model? Will there be so many that any result will be invalidated before it is obtained?

The problems inherent in considering models of animal cells, with their fluid characteristics, has led several research groups involved in modelling developmental systems (Herman in the United States, Lindenmayer in the Netherlands) to concentrate on simulating plant systems. There are fewer examples of amorphously shaped cells in the plant kingdom, and the cellulose-walled, normally rigid plant cell is perhaps the closest biological approximation to the computer array position.

REFERENCES

This chapter considered the general features of and rationale behind computer modelling, looking at how such models may be useful tools in extending our understanding of how organisms develop. The next three chapters look in detail at a range of developmental models using computers, and any reader who has had no previous exposure to computer programming and who wishes to construct his own computer models is recommended to read a suitable computer programming text in parallel to reading the rest of the book. It should be possible to attempt versions of many of the simulations described with only a few weeks' programming experience. Writing bigger and more ambitious programs is the best way to attain proficiency in the art of simulation. Suitable texts are listed in Appendix 2.

More advanced simulation strategies are given in the following two books:

Smith, J. (1968) *Computer Simulation Models*, Griffin, London, UK.

Tocher, K. D. (1963) *The Art of Simulation*, English Universities Press, London, UK.

Chapter 7

Simulating developmental processes: (1) Methods

The sorts of simulation to be discussed in this chapter rely primarily on the spatial arrangement of cells and input and output of information to and from simulated 'neighbouring cells'. The various techniques that have been used in making two-dimensional computer models of developing systems will be listed and described—a more detailed discussion of how such tools can be effectively put together and employed in specific situations follows in the next two chapters. Rather than describe the basic features of cell interaction simulations individually, we will begin by looking at a simple model which uses some of the techniques employed in more advanced systems. Programming terminology will be based on ALGOL.

GROWTH IN A ONE-DIMENSIONAL ROW

The initial basic unit of all our simulations in this chapter will be the cell. It will be represented either by an array element, or group of such elements, in the computer. To represent a row of cells such as an algal filament in computer terms, it is necessary to start by considering a one-dimensional integer array (discussed in the previous chapter). This consists of a variable number of elements that can be referenced in the form,

LIST (1) ... LIST $(N - 1)$ ... LIST (N)

where N is the number of elements in the array. Figure 7.1 shows how these elements might actually appear if the storage locations in the computer that they represent were actually drawn as a series of 'pigeon holes'.

Each location can hold an integer number with a range depending on the computer to be used. We can make each location that we want to represent a 'cell' hold a special number (say, 1) and each location representing 'medium' or a 'space' contain a different number, say 0. Figure

LIST (1) . . . . LIST (N)

| 1 | 0 | 0 | 3 | 2 | 0 | 1 |

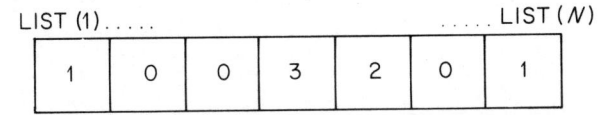

Figure 7.1 A one-dimensional array where $N = 7$. The numeric values of the array elements are shown.

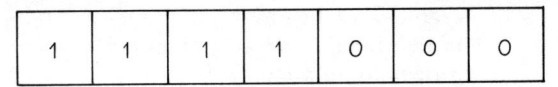

| 1 | 1 | 1 | 1 | 0 | 0 | 0 |

Figure 7.2 An 'algal filament' represented by a one-dimensional array. '1' = cell, '0' = medium.

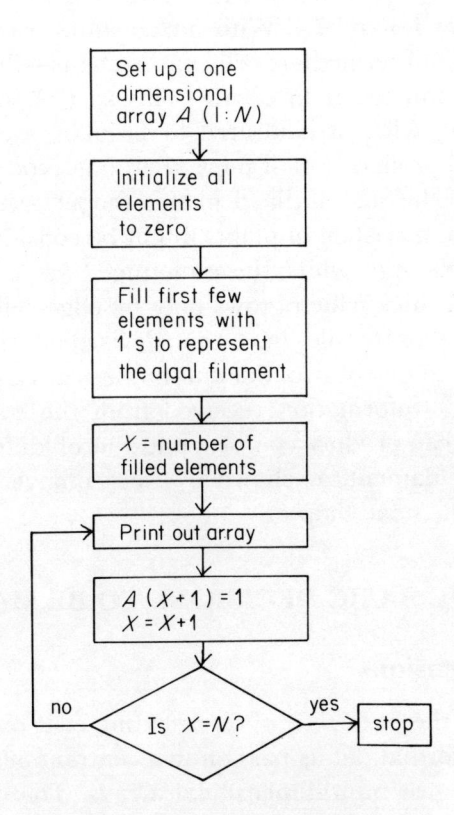

```
┌─────────────────┐
│ Set up a one    │
│ dimensional     │
│ array A (1:N)   │
└────────┬────────┘
         ▼
┌─────────────────┐
│ Initialize all  │
│ elements        │
│ to zero         │
└────────┬────────┘
         ▼
┌─────────────────┐
│ Fill first few  │
│ elements with   │
│ 1's to represent│
│ the algal       │
│ filament        │
└────────┬────────┘
         ▼
┌─────────────────┐
│ X = number of   │
│ filled elements │
└────────┬────────┘
         ▼
┌─────────────────┐
│ Print out array │◄──┐
└────────┬────────┘   │
         ▼            │
┌─────────────────┐   │
│ A (X+1) = 1     │   │
│ X = X+1         │   │
└────────┬────────┘   │
         ▼            │
      ╱      ╲        │
   no╱ Is X=N?╲ yes   │
  ◄─╲         ╱──► stop
      ╲      ╱        │
         │            │
         └────────────┘
```

Figure 7.3 Flow chart for growing the one-dimensional 'algal filament'.

```
1 1 1 0 0 0 0 0 0
1 1 1 1 0 0 0 0 0
1 1 1 1 1 0 0 0 0
1 1 1 1 1 1 0 0 0
1 1 1 1 1 1 1 0 0
1 1 1 1 1 1 1 1 0
1 1 1 1 1 1 1 1 1
```

Figure 7.4 Printout of one-dimensional algal filament growth simulation.

7.2 shows a short, idealized 'algal filament' represented in these terms. To 'grow' this filament, it is only necessary to replace each successive 0 by a 1. The flow chart to do this is shown, together with a specimen printout in Figures 7.3 and 7.4. With such a simple rule, no allowance is made for growth in intermediate cells in the chain—that is, to 'point' at a particular cell and tell it to divide. This sort of sophistication only becomes necessary when it is desired to simulate a cell mass growing into some particular shape, or if more than one type of cell is present. Simple growth of the sort outlined in this model occurs quite often in nature. The apical meristem of plants might be considered a more complex but similar example, whilst the sea lettuce *Ulva* consists of a sheet of cells only one cell thick which grows only by edge-cell division.

In fact, quite elaborate rules have been devised which grow 'two-state' cells which can hold one of two different numerical values depending on 'input conditions' (information received from the surrounding cells). Rules of this type can produce complex patterns of 'differentiation' along the length of the filament, as shown by Lindenmayer, whose work we will consider in the next chapter.

SOME BASIC DECISIONS TO BE MADE

How many dimensions?

Having looked at the basic way of representing cells on a computer in a one-dimensional format, let us pass on to a central question. Computers can deal with one- or multidimensional arrays. The complexity of the program will increase with the number of neighbours that cells have to deal with. In the filament system, each cell has only two nearest neighbours, one to the left and the other to the right. In a three-dimensional model, this figure can rise as high as 27. The complexity involved in writing three-dimensional programs has meant that most cell simulations have been carried out in two dimensions only. A two-dimensional simulation of a three-dimensional system can often give useful information so long as the experimenter is not trying to produce an absolute duplicate of the real system on the computer.

Number of neighbours to each cell

The number of nearest neighbours that each cell may possess in a rigid array (called the 'neighbourhood space') is limited. In a two-dimensional array, there are several configurations which may be used.

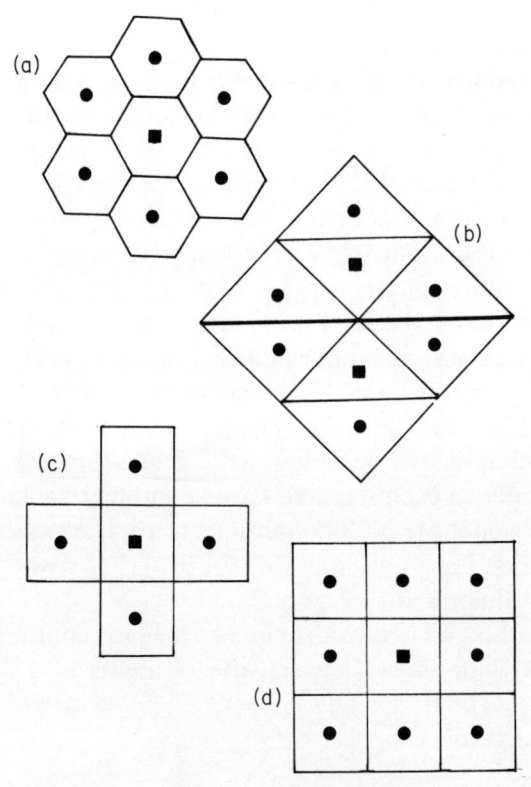

Figure 7.5 Types of cell packing in two dimensions: (a) hexagonal, (b) triangular, (c) square, (d) eight-neighbour. The nearest neighbours of cells shown as solid squares are marked with a solid circle.

It is normally taken (for example, by Williams and Bjerknes, 1972; Antonelli, Rogers, and Willard, 1973) that cells tend on average to pack into a hexagonal alignment *in vivo*. Figure 7.5(a) shows that each unit of a two-dimensional hexagonal array has six nearest neighbours.

There are two other *regular tessellations* of a two-dimensional array: by triangles (Figure 7.5(b)) and by squares (Figure 7.5(c)). An eight-neighbour space can also be defined (Figure 7.5(d))—in this case, diagonal elements are also taken to be nearest neighbours. This produces an *irregular tessellation*, with the centre of the diagonal elements further from the centre of the central element in the ratio 1 : 2. However, this will balance out for all elements in the pattern, and some authors (e.g., Leith and Goel, 1971) have used this arrangement.

Two dimensions out of one

Does a two-dimensional model have to use a two-dimensional array? Two-dimensional arrays are a feature of most computer languages and language implementations where large machines are in use. Computers usually process two-dimensional data more slowly than the equivalent data in a one-dimensional format because they first have to reduce the information to one dimension, and many small machines do not have the facility for handling data in two-dimensional form. Two-dimensional arrays are also cumbersome, because of the need for referencing both an X and a Y coordinate whenever an array position is to be processed. To make two-dimensional techniques more accessible, and to speed up the execution of programs, it is quite possible to treat one-dimensional arrays, or 'lists', as they may be called, as if they are two-dimensional. This is done simply by 'chopping' a long one-dimensional list into a number of segments, each segment corresponding to a successive row of the two-dimensional array. The number of each row is, therefore, the next increment on the Y axis, and the length of each row the maximum X figure. This is illustrated in Figure 7.6.

The relationships between elements of a hexagonal array are rather more complex than those between the elements of a four- or eight-neighbour array. Thus, for any element of a four-neighbour structure, these neighbours are

$$E - 1, \quad E + 1, \quad E - N, \quad E + N$$

where E = the element under consideration and N = the number of elements in a row. The neighbours of E in a hexagonal array also depend on whether the row is inset or not (see Figure 7.7). If it is inset, the

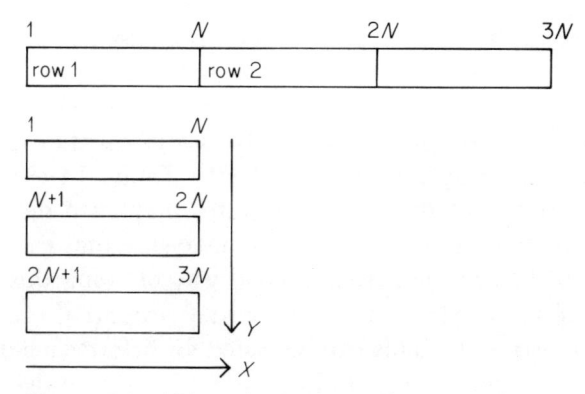

Figure 7.6 'Chopping' a one-dimensional list into a two-dimensional array. N = number of elements per row of the latter. X and Y show X and Y axes of the two-dimensional array.

neighbours are
$$E - N, \quad E - (N - 1), \quad E - 1, \quad E + 1, \quad E + (N - 1), \quad E + N$$
and if not, they are
$$E - (N + 1), \quad E - N, \quad E - 1, \quad E + 1, \quad E + N, \quad E + (N + 1)$$

The edges of the array

A two-dimensional array normally has a square or hexagonal neighbourhood space, and cells dividing on the array will sooner or later

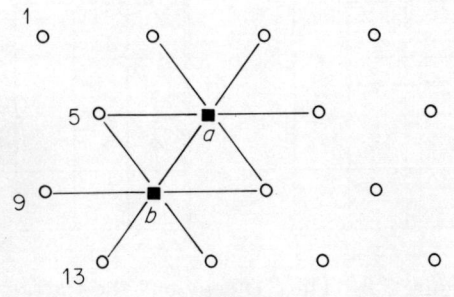

Figure 7.7 Relationship between elements of a one-dimensional array (16 elements). The text describes how neighbours are calculated. Cells a and b have their neighbours indicated.

encounter one of the array edges. If the array used is a *bona fide* two-dimensional array and no safeguards have been built into the program, a non-existent array element may be called up. If a one-dimensional list is used to model the two-dimensional array (see last section), elements on other rows of the array may be called mistakenly. In either case, the program will run wrongly or the computer may even default the program. To stop faulting, we must test to make sure that each array position named in the routine exists. A good way of doing this is to fill the elements in a single element width border around the array with a dummy value (say, -1). This can be tested for before each division by a simple program modification and the site can be ignored if the array boundary is reached by trying to divide in a particular direction. A situation like this might be handled by the programmer telling the simulation to stop at the point the border is reached.

Other ways of handling edges exist. It is possible to fold the edges of a two-dimensional array onto a torus (Williams and Bjerknes, 1972): this means that the effects of reaching a finite edge do not exist. In some computer installations it may be possible simply to fault-trap for illegal array bounds. The 'border' may even have to be more than one cell diameter in width if long-range cell interactions are allowed to take place in a model, otherwise cells may still try to 'interact' with nonexistent array positions, even when a border is present.

In most cases, border cells will behave differently to cells in the centre of the array. Figure 7.8 shows an effect of the boundary on a cell which should move randomly. The effect of the border will increase drastically

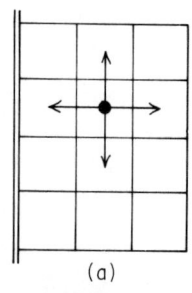

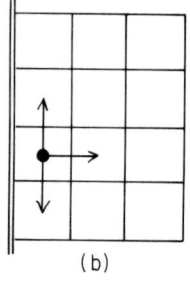

Figure 7.8 The effect of the array boundary on dividing cells. The cell in (a) is surrounded by unoccupied array elements and can move or divide in any direction. The cell in (b) cannot move through the boundary at the left.

Table 7.1 Relative boundary effects for different sizes of square array.

Array size	Number of edge cells	'Edge fraction'
25	16	0.64
100	36	0.36
400	76	0.19
1 600	156	0.10
6 400	316	0.05
10 000	396	0.04

as smaller arrays are considered, as the above table shows. In a square array the number of edge cells is $4n - 4$, where the array size is n^2.

There are two main solutions to minimize the effect of the 'edge fraction'—the proportion of cells on the edge of the array. The first is to use an array of a size sufficient to lower the edge effects. A better method for cell-growth models is to use an array larger than the maximum size that the cell population will grow to. A third method, constraining growth of cells by means of a computer-generated 'membrane', can also be used in special cases, and a description of this method will be found later in the chapter.

A TWO-DIMENSIONAL CELL POPULATION GROWTH MODEL

The first models devised to simulate two-dimensional cell population growth and morphogenesis were elementary in structure, and models of this type can be used to illustrate the one-dimensional 'chopping' technique and the use of array elements in a 'square' four-neighbour format. The initial step is to choose a suitable array size, say 1600 integer units, which will equal a 40×40 two-dimensional array. In the computer language ALGOL, a suitable one-dimensional array might be defined as $M(1:1600)$. The array is then initialized with all elements set to contain the 'medium' or 'empty' value of 0. Next, a 'starter cell' is placed in the centre of the rectangle of the array (this will be at about position 820). The number '1' is used to represent a cell, so $M(820)$ will now equal 1. The next requirement is a set of growth rules to enable this solitary cell to divide.

Cell division

As this is a 'square' growth model, the cell can divide in any one of four directions. When only one cell is present, there is only one choice of the next cell to divide. The cell to replicate when many cells are present is chosen by using a 'random picking' method: this technique will be described below. Division takes place by making one of the neighbouring array sites equal in value to the 'dividing' cell. Hence the original site becomes one of the two 'daughter' cells. In computer terms an assignment statement such as $A(K) = A(L)$, where L is the array coordinate of the dividing cell, and K the coordinate of the space this cell is dividing into, is used to simulate the replication process.

In the elementary model, we will assume that only cells with at least one free space in the immediate neighbourhood can divide, thus avoiding any worry about having to 'push' neighbouring cells out of the way to create space for a new cell to move into. The following example (Figure 7.9(a)) shows how a section of program can search for a space around a chosen cell, and allow the cell to divide randomly into one of the available spaces. $RIR(X, Y)$ is the name of a function returning an integer random number in the range X, Y. A program written on these lines will multiply from the initial cell until the array boundary is exceeded (and the program is faulted). Figure 7.9(b) shows what happens in a particular example of this kind.

Control lists

Unless all cells in a model system always have an equal chance to divide, there has to be some ordering process controlling which cells are able to divide at any time. This is done by using a *control list*, also sometimes called a *hit vector*. The most convenient time scale is the generation, defined as the period in which all cells have been allowed to divide once. A control list is a convenient method by which the division state of all the cells in the model can be determined. In the present model, the list must also include facilities for removing cells from the list when they can no longer divide (that is, when they have become surrounded by other cells). In more sophisticated models where all cells can divide all the time, this step is not necessary. The control list is set up as follows.

A one-dimensional array is maintained, with each element storing the location of one of the 'cells' on the main array. At the start of any generation, the list will need to be twice as long as its length at the end of the previous generation, to allow for the growth that will occur during

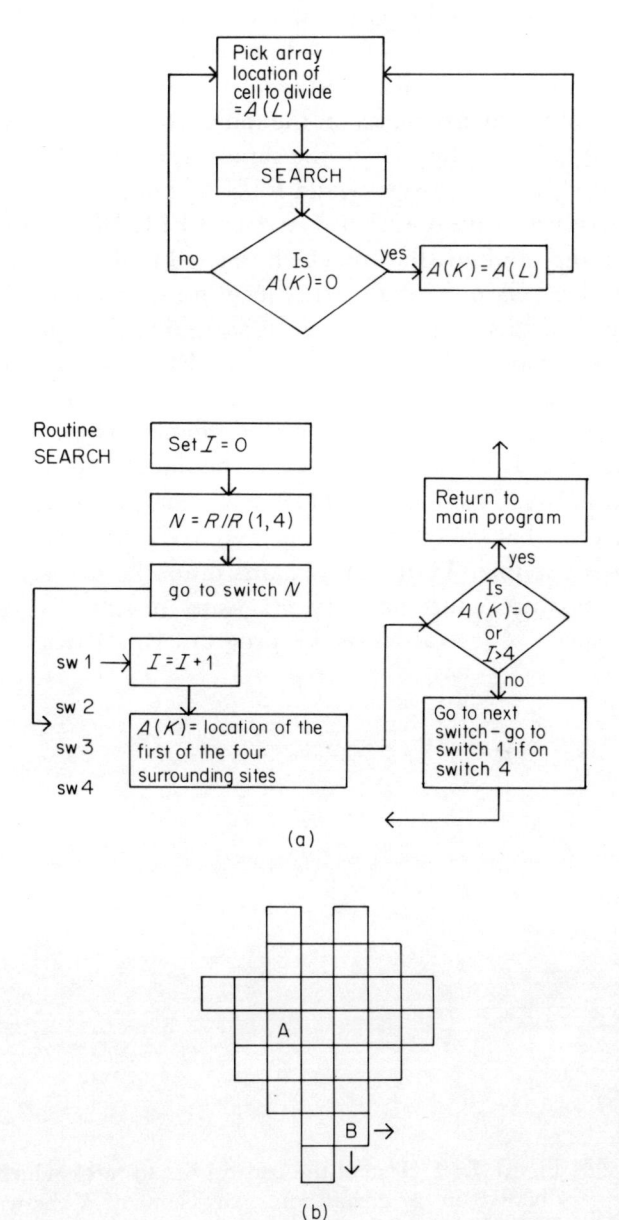

Figure 7.9 (a) Flow chart for searching neighbouring array locations to see if a space is available to divide into. Only the first of the four possible search directions is shown programmed in full. (b) Cell A cannot divide. Cell B can divide in two possible directions.

the new generation. The list is set up as follows (Figure 7.10). At each new generation, the number of 'cells' present becomes equal to a variable we will call TOP. The list locations 1 to TOP then become filled by the locations of the array points on the main array that contain cells. A second variable FLOAT is set to the same value as TOP.

All the list elements between 1 and FLOAT are then processed in the following manner. A number between 1 and FLOAT is chosen at random (this is the random step by which the cell to divide was chosen in Figure 7.9). The cell in the main array location stored in the list element chosen then attempts to divide. If this is satisfactorily done, then all the list elements between the one picked and FLOAT inclusively move down the list one position, so removing the chosen element. FLOAT is then decreased by one, and the processed element is placed in the old FLOAT position. If the cell has divided, the daughter cell has its array coordinates stored in list location TOP + N, where $N - 1$ is the number of daughter cells that have already appeared in the present generation. If no division is possible, TOP + N remains unused.

Although this procedure has advantages in 'neatness' it can only be used when speed of operation of the program is not essential. This is

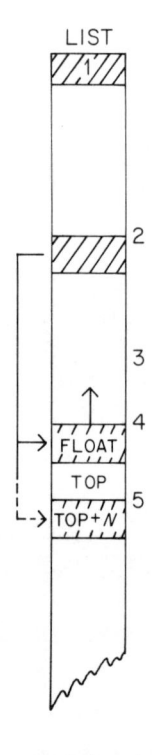

Figure 7.10 Use of the control list to track which cells have divided in a generation. An element N between 1 and FLOAT is picked at random. The main array element stored in LIST(N) is then divided. FLOAT is decreased by 1 increment. The divided cell's main array location is placed into the old value of FLOAT and the new cell's main array location is placed in LIST(TOP+N) where N is the number of cells divided in that generation. At the end of the generation FLOAT is reset to TOP+N.

Figure 7.11 Random division in a mass of computer 'cells'.

because for every division, an average of one-half of the cells which were present at the start of the generation will have to be moved down the list by one position. A quicker way of carrying out the 'cleaning up' procedure, by which dividing cells are placed at the bottom of the list, would be to let, say, 10 cells divide. These cells are then processed in the same way as before except that their original positions are marked by some dummy variable, for example '9999'. After the given number of divisions, the whole array is swept through once, and all '9999' positions are excised. In this way, about nine-tenths of the laborious array position changing inherent in the first variation can be avoided. If a '9999' is hit during random casting for cells to divide, a new cell is picked to divide. A typical printout of the whole cell growth simulation after one, three and six generations is shown in Figure 7.11.

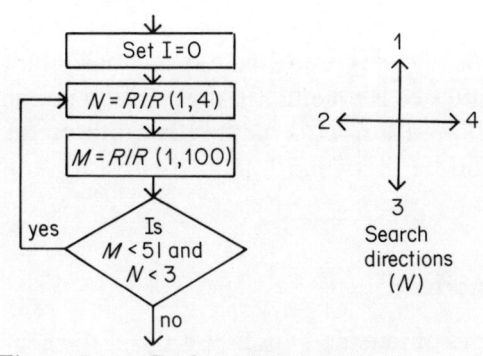

Figure 7.12 Preferential searching rule to increase number of divisions to the right and below the dividing cell. In half of the cases where division directions 1 and 2 are picked, the computer makes a fresh choice of division direction.

Changing parameters

It is only in variations on this simple scheme that the first uses of such computer models become clear. Suppose we ask what happens if the computer cells divide preferentially in certain directions. Preferential division is easily programmed into the model just described. All that is needed is to include an increased probability of searching in a particular division direction. The start of the example in Figure 7.9(a) might then read as in Figure 7.12. This would give a 50 per cent greater probability that searching will take place to the right, and below, the parent cell.

A MODEL THAT ALLOWS ALL CELLS TO DIVIDE

Simplified method

The main limitation of models of the type described above is that only cells with free edges can divide. If only one cell type is to be modelled, an alternative procedure can be used. Cells are picked from the control list in the same way as in Figure 7.10, but instead of being dropped from the list when they become surrounded by other cells, the program finds the 'nearest' free space to the cell to divide, and puts the daughter cell in this space. The free space is found by successively testing the array positions in each of the four directions allowable for division until it locates an empty space. This has the advantage of allowing cells *inside* a cell mass to divide (after a fashion) without recourse to more complex 'pushing algorithms' that would be needed to clear a space for the daughter next to the parent cell. Note that the sole reason we can use this technique is that only one type of cell is modelled: no way of simulating 'cell lineage', by placing daughter cells together in the position previously occupied by the parent, is allowed for. This method is quite useful for less sophisticated models, and has been used to good effect by Ede and Law (1969) (see Chapter 8).

The pushing model

The disadvantages of placing daughter cells on the nearest available free 'edge' of the cell mass have already been mentioned. To overcome them, it is necessary to clear a space next to the cell to be divided. To do this successfully requires care, because moving computer cells involves transferring them from one array location to another and mistakes can cause the program to default. It should be noted that the rigid nature of the array makes it well nigh impossible to carry out division of the sort seen in living organisms (or even in monolayer cultures, the closest real-life equivalent to the two-dimensional simulations).

To illustrate this point, consider a mass of typical amorphously shaped embryonic cells. If one of the central cells in this mass divides, there will be an incremental change in shape of a large portion of the cell mass, as a space is allowed for the new cell. If a computer 'cell' X divides into a space Y, which is already occupied by another cell, and this latter cell is moved, it has to travel a whole array position in distance, which means the same thing has to happen to one of its own neighbours in turn, and so on out to the edge of the cell mass. There is, then, a marked difference in these two situations, and the modelling of cell divisions within a cell mass either has to allow for this 'single cell pushing single cell' phenomenon, or has to model cells as taking up more than one array position, so that 'flexibility' can be introduced. In this latter case a single cell division giving rise to an incremental position change in a number of cells, as seen in living tissues, can be modelled. As direct cell/cell pushing is easier to program, we will look at this technique first.

A central question to ask is, of course, can the 'pushing' type of movement ever be analogized to any growing tissues? The answer to this is best illustrated by a practical example of the use of this kind of model in a specific experimental context. Such a model system is described in Chapter 9, and indicates that the effect of many pushings in a large simulated cell mass may give results not too far removed from the incremental 'jostlings' seen in living organisms.

There are two variations on the pushing theme. In the first, the cell computes the nearest free edge to itself, and all cells in the line between the free space and the cell to divide are moved outwards one position to create a new free space. The daughter cell can then be placed in this space. The most convenient way of programming the successive movements of the intervening cells is to move the outermost cell first—this would be ludicrous in any real division system, but it leads to the same end-product (Figure 7.13(a) and (b)).

An alternative procedure, although more costly in computer time, would be for each cell in turn to compute the nearest free edge, and push itself into the position occupied by its neighbouring cell in this direction. The neighbour, with no real 'position', would have to repeat this procedure. The advantage here is that the path to the edge from the cell that initially divided in the model is rarely a straight line, thus 'slicing' by constantly pushing a regular file of cells is avoided. It must be added here that no practical evidence exists that the latter method produces any noticeable difference in the separation of related cells within a cell mass, although its use is obligatory when using multi-array location cells (see below). Using programs of this type, it is possible to 'tag' cells (by making them of a different numeric value), and to observe 'clone formation' by these tagged cells. The method can be used for any number of different clones, because the daughters always take the numeric values of the parent cells, and can thus be easily kept track of. The importance of clone formation will be discussed in Chapter 9.

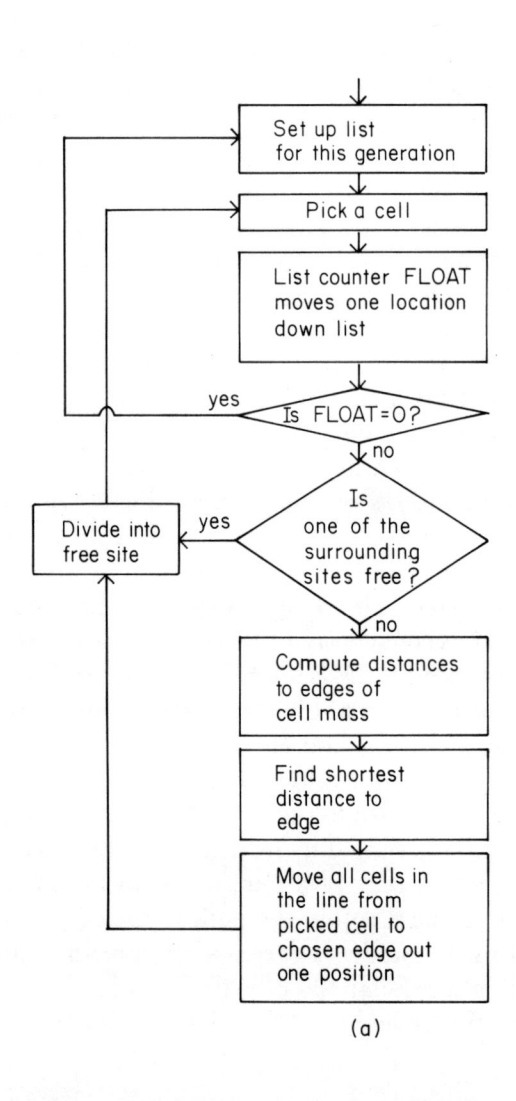

(a)

(b)

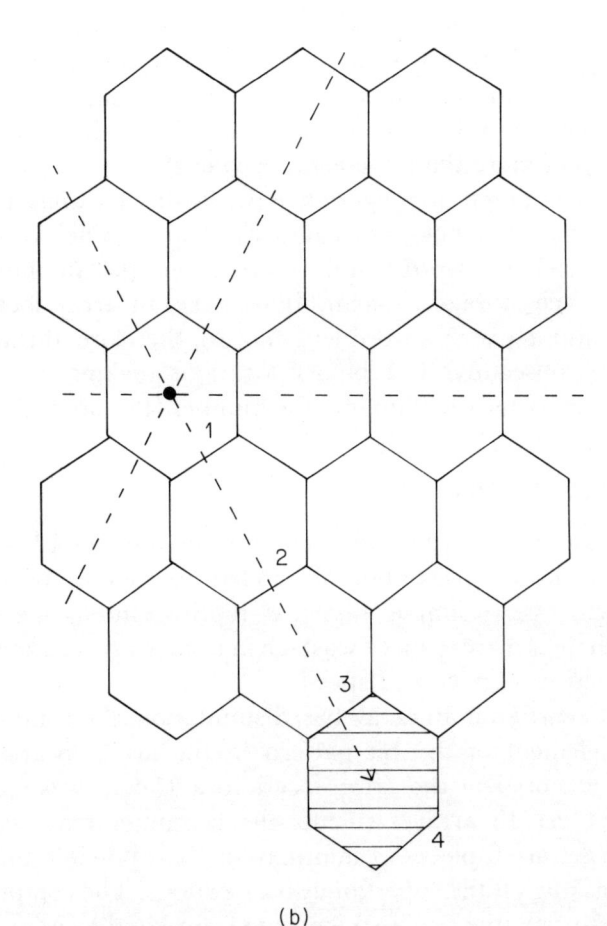

Figure 7.13 The pushing algorithm. (a) Flow chart showing how 'pushing' a line of cells to create a space for a dividing cell is carried out. (b) Diagrammatic representation of the pushing scheme. 'Cell' to divide (1) finds the nearest edge of the colony by adding up the numbers of cells in the possible division directions (six are shown in this example). The cell decides to divide to the bottom right. Cell (3) moves to space (4), cell (2) moves to the space vacated by cell (3). Cell (1) can now divide into the space left by cell (2).

Bit packing and 'complex' cells

Up until now, we have considered the array elements as carrying an 0 or 1, signifying whether or not they contain 'cells'. Each array element in the computer is a computer 'word' that can hold digits up to the word length of the computer being used, and so is capable of holding much more complex data than a single unit. In the following pages, two techniques using this feature will be used. The first is 'bit packing' to get more information into one array element. The second is a logical progression using the technique of bit packaging to build up complex cells made up of more than one array element.

High-level computer languages have the facility for 'masking' certain of the binary locations holding each array element. Each computer has its own characteristic 'word length' representing the number of binary locations in each storage element. If we take an array location on a particular computer with a word length of 16, therefore, this means that there are 16 consecutive 0–1 binary locations making up each array location. An array location holding the number '12', holds the following bit pattern:

000000000000 1100

This is because the binary representation of the number 12 is 1100, and the computer fills its words from right to left. As we have been considering cells with very simple numeric representations—1 or 2, for example—there is a great space wastage in each array location. This has two consequences—space and speed.

In order to carry out an array-based simulation, the computer has to scan every element of the bit pattern in the array locations. If the number '1' is stored in the array location, a 15-fold waste of effort is introduced. Over 16 array elements the computer may scan 16^2 bit locations to get at 16 pieces of information. It is possible to avoid this pitfall by masking off the superfluous bit locations. The computer would then only scan, say, the one, two, or three right-hand locations, depending on the simulation. This masking facility can not only help with processing array elements more speedily. Any combination of array elements can be masked off at any time, so that a number of 'cells' can be held in the same array element. Conversely, a cell might be split into a number of parts, which can all be referenced in the same array element.

The disadvantages of modelling cells as one inflexible unit have already been considered. It is possible to surmount this difficulty by using more complex cells and division procedures. Complex cells may, of course, be made up of any number of array positions greater than one; the easiest to describe and program contains three array positions per mature cell. A model of this type can even include facility for cell growth before replication—cells do not suddenly double in size, but gradually increase in size from two to three array positions before 'splitting' into two daughter cells. The previous models described in this chapter have not taken this into account.

Picture a hexagonal array, with any two array positions 'joined', as in Figure 7.14. Such a pair of array elements could be said to make up the basic cell, and the cell increases in size by one of its halves 'duplicating' (Figure 7.14(b)). A further duplication then leads to splitting into two daughter cells. A variety of configurations can be produced by two or three array element cells, as shown (Figure 7.14(c)). The most useful characteristic of cells like these is that they can now be 'pushed' without

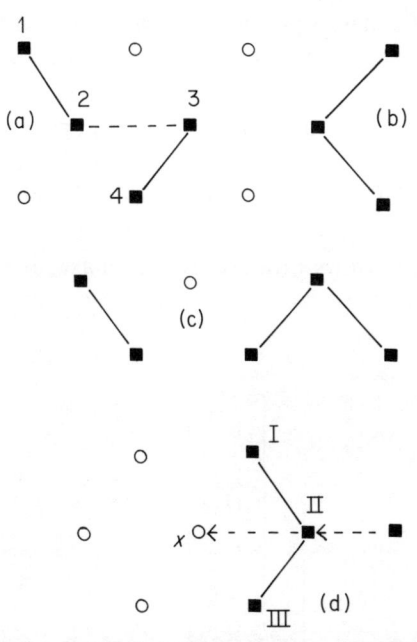

Figure 7.14 Configurations taken by cells occupying more than one array position. Solid squares represent parts of cells. Open circles represent empty array positions. (a) Two cells (1,2 and 3,4) which have recently formed. The dotted line shows how the original cell (elements 1,2) first grew to three parts (1, 2, 3). (b) A three-part cell. (c) Different configurations shown by cells. (d) Partial movement of a cell. The element on the right-hand side pushes the central (pivot) element of cell (I, II, III) to array element x.

having to move their whole mass, so modelling the shape changes seen in real cells (Figure 7.14(d)). To model cells like this using the conventional method of one array element per cell method would be most wasteful in terms of computer time and space. If parts of cells can be 'squeezed' into the same array location, simulations can be run much more efficiently.

The data referencing the complex cells is held in the computer as follows. The cells in the model are initially made up of two array elements, as shown above, and these are held in the computer in the same format as we have discussed previously—a main array and a control list. The list is rather more involved in this case, however—three separate pieces of information are compressed into a single word (see Figure 7.15). The right-hand section carries the main array coordinate of one of the component cells, and the centre section holds the coordinate of its partner. The third (left-hand) section is initially left unfilled. When the cell 'grows', the new part of the cell takes the fresh list element (TOP + N), as in previous models. In this case, it occupies the centre

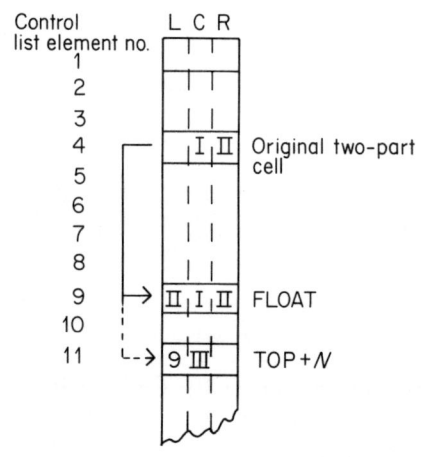

Figure 7.15 The control list used for two- and three-array position cells. Each control list element location holds three pieces of information. List element 4 is a two-array position cell, where two parts of the list element hold the main array locations of the cell parts. When it is picked to grow, list element Top + N holds the location of the new part of the cell in part C. Part L holds the list location of the other two parts of the cell, which are put into FLOAT. Conversely, part L of the list element Top + N holds the list location of the third cell element. Cell division occurs by the third cell being picked to divide, in which case the linkage relations (in L) between the two two part cells formed are wiped out.

part of the word. The parent list element takes the old FLOAT position. The left-hand parts of both words now take the value of the *list position* of the partner; the daughter takes that of its parent, vice versa, for the parent location.

To carry out 'growth', one of the two parent units has to divide in much the same way as in the examples of single array position cells discussed above. The parent unit in question now becomes a so-called *pivot* unit, for the following reason. If a cell divides or grows in the centre of the cell mass, it has to push itself a space. If all cells are one array unit only in size, there is no problem—cells in this type of model can move about freely. With complex cells, it may be that part of the cell may be physically separated from the rest of the cell if this rule is not modified, and so one of the units of the cell is designated the 'pivot'. The pivot unit can be moved only to positions where it is still in contact with both the other units of the cell. In Figure 7.14(d) for example, the pivot unit (II) can only move to the site marked X. The other units of the cell are free to be moved independently of each other, so long as they remain in contact with the pivot unit.

The pushing rules used with this model are a modification of the single array position model where individual cells have to find the nearest edge, and then dislodge neighbours sequentially out to this edge. It is impossible to use directly the 'slicing' technique (p. 95) to clear a space, because there might be pivot cells in the line that cannot move out in the right direction. A modification of the space clearing algorithm allowing for 'detours' therefore has to be used. A word of caution—this type of model is slow in operation, because of the many list-element retrievals that have to be made. Also, using the program to look at clone growth shows no real advantage over the less costly, simple pushing model.

CONSTRAINING GROWTH

We have already discussed how the presence of a 'border' can prevent cells dividing in particular directions. This effect can be modified from its passive role, and may be used to model cell population growth under physical constraints simulating, say, membranes or surrounding tissues, both of which have been shown to be important during embryogenesis. The simplest kind of 'physical constraint' is the prevention or limitation of cell growth in particular directions, and this is analogous, to a certain extent, with the external forces provided in real biological systems by membranes. In this case, it is possible to restrict cell division in certain directions merely by modifying the 'random cell division' algorithm shown in Figure 6.1. However, it may be that a specially shaped or growing membrane is to be simulated, and this will require more sophisticated programming procedures.

An ideal way of simulating a growing constraint is to embed a formula defining a geometric shape, like a circle or an ellipse, directly into the program. It is then possible to repeatedly increase the size of the constraint and follow the growth of a cell mass growing inside it. An example of a program of this sort will be discussed in Chapter 9.

Constraints are modelled by using a system of specially coded array elements as in Figure 7.16. A circular constraint would have a radius R. All array elements this distance away from a series of chosen central array elements would be coded as 'constraint' locations. If a cell tries to divide into one of these locations it is prevented in much the same way as are edge cells (Figure 7.8). To change the size of the constrained space (to mimic a 'growing' membrane, for example), a new radius is calculated. Array elements on the circumference of the newly calculated circle

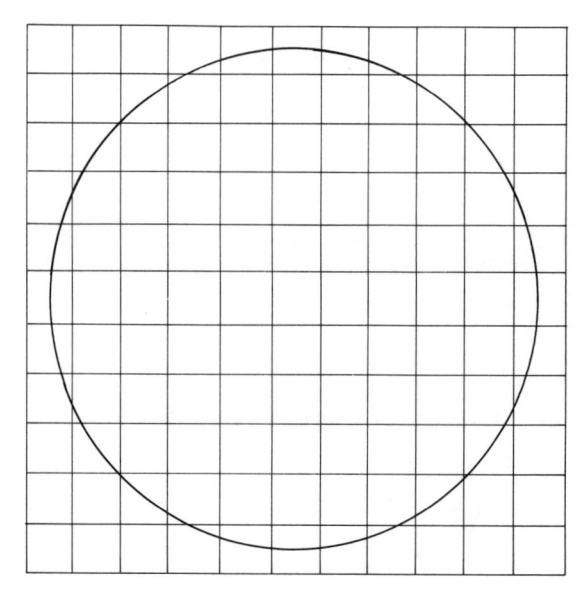

Figure 7.16 Specification of a circular constraint 'membrane' on a two-dimensional array. Each square represents an array element: squares containing the circular constraint define the constraint, and only array locations *inside* the constraint are available for cell division.

are coded accordingly and the previous circumferential array elements are made available for cell division.

A TOPOLOGICAL MODEL OF CELL INTERACTIONS

The array structure for simulating cell systems has a number of disadvantages associated with the rigidity of the array. We have already looked at a mechanism for simulating 'complex' deformable cells on the computer, but the increase in program complexity even for cells with two or three 'parts' is immense and the question must be asked, is it worth a significant amount of effort to simulate what is essentially a modest improvement on the basic single array element model?

A recent advance by Matela (Matela and Fletterick, 1979, 1980) is a novel approach to avoid the pitfalls of array-bound cell modelling. The technique used involves *graph theory*. Instead of being restricted to a rigid

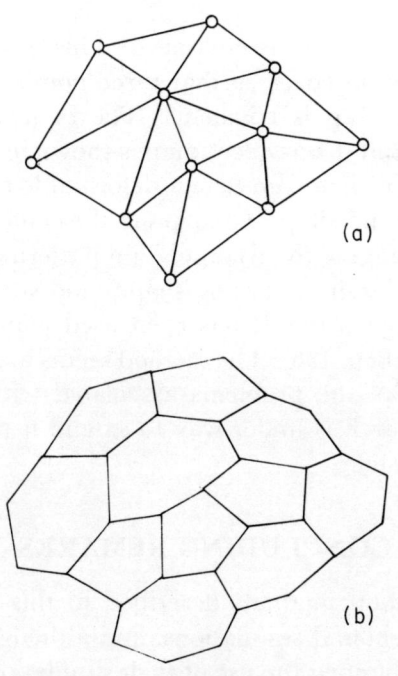

Figure 7.17 Topological representation of a cell system. The triangulated graph (a) is a representation of the geometry of the cell system (b).

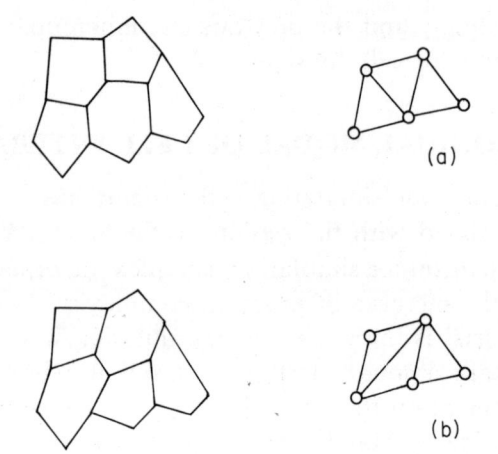

Figure 7.18 Changes in cell conformation by means of local exchanges in the triangulation pattern of the graph.

lattice, a triangulated graph is envisaged. This is simply a number of points ('cells') joined in space so that three points are incident to any given junction (the map is trivalent). The equivalence between the triangulated graph and the trivalent map is shown in Figure 7.17. As can be seen, the net result is to simulate a deformable cell mass which has cells of variable size and shape. Changes in the conformation of the cells is carried out by changing the triangulation pattern on the map (Figure 7.18). This has the result of varying shapes and sizes of the cells.

As yet the topological model has been used mainly to simulate cell sorting (see Chapter 8, p. 136). The method seems to be powerful enough to overcome many of the problems associated with array-bound cell modelling, and research is under way to extend it to simulate growing and dividing cell systems.

CONCLUDING REMARKS

Most of the simulation methods described in this chapter have been developed for array-bound simulations, and although the previous section seems to have damned the use of such simulations there are certain advantages inherent in their use. The major advantage is programming simplicity: it is easy to reference cells spatially using an array format. An additional factor is the spatial relationship between the array locations in the computer and the cell mass as printed out. Topological models have no such direct reference between the program representation and the real pattern on a print out, and the geometric representation of topological cell models is as yet an unresolved problem.

This chapter has dealt with the bare bones of the simulation methods. In the following chapter, particular model types will be described to give the reader an idea of the different techniques used in modelling developing systems.

REFERENCES

There are few accounts that specifically deal with the methods used in computer modelling in developmental biology. The technical sections of many of the papers to be discussed in the next chapter will probably be found most useful (see reference guide to that chapter). Of particular interest to the array-bound modeller are the following papers.

Antonelli, P. L., Rogers, T. D., and Willard, M. A. (1973) Geometry and the exchange principle in cell aggregation kinetics, *Journal of Theoretical Biology*, **41**, 1–21.

Ede, D. A., and Law, J. T. (1969) Computer simulation of vertebrate limb morphogensis, *Nature*, **221**, 244–248.

Leith, A. G., and Goel, N. S. (1971) Simulation of movement of cells during self sorting, *Journal of Theoretical Biology*, **33**, 171–188.

Matela, R. J., and Fletterick, R. J. (1979) A topological exchange model for cell self sorting, *Journal of Theoretical Biology*, **76**, 403–414.

Matela, R. J., and Fletterick, R. J. (1980) Computer simulation of cellular self sorting: a topological exchange model, *Journal of Theoretical Biology*, **84**, 673–690.

Ransom, R. (1977) A computer model of cell clone growth, *Simulation*, **28**, 189–192.

Williams, T., and Bjerknes, R. (1972) Stochastic model for abnormal clone spread through epithelial basal layer, *Nature*, **236**, 19–21.

Chapter 8

Simulating developmental processes: (2) Models

INTRODUCTION

There have been few reviews written up until now on computer models in developmental biology, and this chapter will try to give an account of some of the models used by other workers over recent years. It does not, of course, claim to be a complete list, but examples of most types of computer model used are included. The authors own models are discussed at more length in the next chapter. Other accounts of developmental models include that of Rosen (1968), who deals with general cellular control processes, and theoretical models made of them; he is not concerned with development as such, although he mentions several of the examples to be discussed below. In a second paper (Rosen, 1972), the same author reviews the theoretical aspects of morphogenesis. This review is largely limited, however, to an appraisal of 'self-organization' phenomena—molecules, viruses, and at a higher level, cellular re-aggregation. Apart from Rosen's work, discussion of various computer models is to be found in Apter's *Cybernetics and Development* (1966). This book is now rather dated (the majority of models to be described in the present chapter are post-1966), and as the title suggests, is biased towards the cybernetic approach: trying to show the validity of considering development within the reference points of cybernetics, as laid out in Norbert Weiner's *Science of Control and Communication in the Animal and Machine* (Weiner, 1948).

A review of computer models in developmental biology may be written in several forms. It is possible to describe the literature 'system by system', saying what models have been made of which biological systems. This does not work out very well because there is no logical order in which to fit such a grouping; the subjects chosen for simulation have been very diverse. It is also possible to list the various models historically, from the first models of von Neumann and Eden in the 1950s onwards. As a third possibility, the simulations could be described from a computer viewpoint—the main feature of model X is the use of list processing, model Y, video techniques, . . .). Because all these methods have their own faults, the models described in the present chapter are placed in groups according to the biological *level* which they simulate.

Organisms are made up of cells, and the first models to be discussed are those which simulate the workings of individual cells. From this relatively simple beginning we move up to the level of cell–cell interactions, starting with the most abstract models constructed: those based directly on automata theory. Most models in this category simulate very general aspects of cell–cell interactions. The next section deals with cell interaction models applied to specific systems, starting with the simplest organisms and going right up to problems of cancer formation and the development of the vertebrate limb. The chapter finishes with a discussion of pattern specification models. You may recall from Chapter 3 that pattern specification mechanisms are the processes by which cells receive the specific information which they need to differentiate: they provide the 'map' by which cells work out where they are and what they are to do in a developing system. Models in this category are less reliant on parameters like cell movement and neighbourhood spaces. Some such models do not even model cells at all, and are more concerned with the way in which particular patterns—for example, those that appear on a sea shell—are set up in space.

SUBCELLULAR MODELS

Relatively little work has been done on the simulation of subcellular development. There is a fairly large body of literature dealing with continuous mathematical models of enzyme fluctuations in particular systems (a continuous model may be defined as a model which involves the solution of mathematical equations rather than using the algorithmic approach set out in the last chapter). A typical example is given by the work of Ycas and coworkers (Ycas, Sugita and Bensam, 1965).

Wright (1973) expounds the basic creed of the enzyme modelling approach: that we must find the biochemical controlling factors of a developmental pathway in order to be able to understand it. The 'critical variables', according to Wright, are invariably measurable concentrations of cellular components. A similar approach is seen in the work of Heinmets (1970). Heinmets uses the analogue computer to process systems of differential equations representing 'enzyme synthesis', 'cellular growth', and 'reproduction'. Heinmets considers two main systems. The first is a single-cell model with an enzyme system which can sense the cell's surroundings. The second model deals with the kinetics of cell interactions such as those leading to recognition and the formation of specific cell attachments in cell aggregates. This model uses some of the properties of the first model—'sensing' exoenzyme elements associated with components transporting reaction products into the cell, where their action can lead to induction or repression.

Both Heinmets and Wright believe that living processes are *complicated* but not *complex*: that these systems provide only problems of magnitude and not of concept; they suggest that the computer simulation of non-linear problems, with appropriate variations of different parameter values fed into the 'equations' describing the modelled system, can provide the key to understanding the problems of cellular biology.

Fluctuations in enzyme level often occur as a direct result of gene control, and models have been proposed to simulate these cellular control processes. Stahl (1967) suggests that cellular control might be represented in terms of the logical manipulation of nucleic DNA and RNA molecules. Stahl first proposed the idea of using a computer-simulated Turing machine[*] to do this in 1963, and the subsequent development of list-processing techniques[†] allowed modelling of a 'complex' cell system

---

[*] In a physical form, a Turing machine may be represented as a roll of paper tape with a scanner. A finite automaton, such as we are considering, possesses a tape that has a finite length, and we define the machine in terms of tape symbols $S_0, \ldots, S_n$, where each subscripted S represents a tape square on the machine; and state symbols $q_0, \ldots, q_n$. A quadruple is a collection of four symbols, specifying the program for the machine, and making the scanner move backwards and forwards over the tape in a prescribed manner. The tape is ruled into squares, each square having one tape symbol only

e.g., $q_0 S_0 S_1 q_1$

means that 'when the machine is in state q_0 scanning square S_0, it must move into square S_1 and state q_1'. Also,

$q_0 S_0 \to q_1$

indicates that the machine is in state q_0, square S_0, moves one square to the right, and takes up state q_1. A Turing machine can be used in this way to handle numbers and functions as a *decision procedure*. It is, in fact, a primitive computing mechanism.

[†] List-processing languages have several important features:
  (i) The program involves manipulating symbols that have non-numeric meaning.
  (ii) The storage requirements in the computer need not be specified in advance; complex data structures are developed as the program is carried out.
  (iii) The relationships amongst the elements of the data are restructured whilst the program is operating.
  (iv) The program can be described at several levels of detail, and the problem is stated naturally in hierarchical fashion.

consisting of 46 genes that could demonstrate repeated 'self-reproduction' and 'differentiation'. DNA coding for each enzyme is represented by a four-digit identification number, and these numbers are held in a list. The RNA list contains complementary copies of the DNA, these copies being generated by an automaton acting as 'RNA polymerase'. An 'enzyme list' contains the 'protein equivalents' (generated by the 'ribosome automaton' of the cell) corresponding to the RNA coding. Self-reproduction involves a series of algorithms representing a build-up of the substrates that the 'cell' needs. In Stahl's model, this is followed by the temporary cessation of activity of most genes, activation of 'DNA polymerase', and strand-separation simulation to give two 'cells'.

A special feature of the model is the 'differentiation controller' gene system which works as follows. Self-reproduction continues, providing the parent cells are still allowed to divide. Once a small colony of 'cells' has been formed, a build-up of by-products triggers the action of a differentiation initiator game, stopping further reproduction, and causing the differentiation of cells into 'inner' and 'outer' types.

These list representations of cell components could theoretically be used to list any DNA, RNA, or protein sequence. The substances used are arbitrarily simplified, in order to be simply handled by the computer program, but a similar methodology could be used to list, say, triplet codes. Stahl's procedure uses discrete, not continuous, techniques and the main operations are integer threshold checks, additions, and subtractions. It is suggested (Stahl, 1967) that program subroutines could be built in to simulate various forms of enzyme kinetics to help avoid the inaccuracy problems inherent in discrete programs.

With Stahl's models we impinge on the field of automata theory, and this tends to pervade the whole literature of the computer simulation of development. It is not difficult to imagine the strings of polymer molecules in cells as abstracted strings of symbols printed on the tape of a Turing machine, with the enzymes governing the rules of manipulation. This treatment of subcellular processes has been criticized by Goodwin (1970), who thinks that the discrete treatment is too far removed from the real situation. The important points in any developmental process are the switching points, say Goodwin. Where cells must decide which of several alternate paths of differentiation to follow they must, in the language of computer programming, obey some subroutine which comprises a form of conditional instruction: 'If condition A is satisfied, do B.' This indicates that if a cell works in the same way, it 'computes its own state, looks at the DNA program, and then changes state accordingly'. Goodwin argues that at the molecular level we will have problems with biochemical 'algorithms' because 'biochemical switching is only analysable in terms of a (non-linear) interaction between the variables describing the state of the cell'.

'let us suppose that a particular metabolite, y, is formed from two precursor metabolite molecules, u and v, the reaction being catalysed by an enzyme E_1. Suppose that a second metabolite, z, is formed from the precursors v and w, E_2 being the enzyme catalysing this reaction. If u, v, and E_1 are present in sufficient concentration but either w, or E_2 is absent, then y will be formed. If v, w, and E_2 are present in sufficient concentration, but either u or E_1 is absent, then z will be formed. These statements constitute the biochemical form of the conditional sub-routine' (Goodwin, 1970).

Goodwin stresses that, although this is clearly what happens, there is no 'Independent computation of cell state, reference to DNA for instructions, and subsequent change of state in accordance with these instructions', as an automata theorist might suggest. The role of the DNA is passive in that the state of the cell at any time is partly determined by the regulation of rates of molecular synthesis by DNA control.

It might be argued that there are some state changes which do appear to be directly mediated by DNA reference, for example control of induction and repression of gene expression in bacteria. In such cases, the computer analogy may hold. Here, Goodwin's argument that 'decisions are taken by the phenotype,* whose properties are largely determined by the genotype†' may be true.

Lindenmayer (1971) has defended the position of the automata theorist in subcellular biology. His argument is as follows. It is quite possible to show the algorithmic operation of the DNA program. For example, the set of all the metabolites in the cell excluding active proteins and nucleic acids may be called C. At any moment in a given cell, a combination of the elements of C (a subset) will exist. If we look at each gene and the enzyme it gives rise to as a *transformation rule* (transforming certain elements of C into other components), we can call the set of all these transformation rules P. At any time in a given cell, a subset of P will be present. We can, therefore, assign a state consisting of subsets of P and C

Phenotype: the physical characteristics manifested by an organism.
†*Genotype*: the 'genetic' constitution of an organism. It is possible to have organisms with the same genotype, but different phenotypes (owing to environmentally produced variation).

to the cell. The *inputs* and *outputs* consist of fluctuations of C in and out of the cell. The *next state* depends on the cell's present state and the inputs it receives. Lindenmayer states that the only real objection against these kinds of constructs in subcellular biology is an objection against their discreteness, as opposed to Goodwin's argument as to their conceptual value. Lindenmayer stresses that the advantages of dealing with integer entities include conceptual and computational simplifications which may be necessary before we can understand these systems.

AUTOMATA-THEORETIC APPROACHES

Early work

The initial usage of automata theory in a biological context comes from the work of John von Neumann two decades ago (von Neumann, 1966). Von Neumann's main interest was in self-replication, and he worked on two models of self-replicating systems. The first of these was the kinematic model, a hypothetical robotic representation of a self-replicating system. It consisted of a central set of computing elements—switches and delays, controlling an artificial hand, or muscle-like element; cutting and fusing elements to make new parts; rigid elements out of which the other types were made; and sensing elements. The automaton 'sat' on a lake of parts, and made copies of itself by following Turing machine-like instructions, sensing and scavenging for new parts.

The problems behind making a workable kinematic automaton would have been immense, and von Neumann soon switched to the two-dimensional 'cellular' automaton. In this way, we are introduced to the concept of the 'two-dimensional array' as a coordinate base for developmental models: these arrays are an essential part of many of the simulations to be described here. The notion of a cellular automaton is built up in the following way and uses many of the concepts introduced in the last chapter. We begin with a cellular space, consisting of an infinite n-dimensional Euclidean space, together with a neighbourhood relation defined on this space. The neighourhood relation gives the list of neighbours for each cell. There is also a discrete time base $t = 1, 2, 3, \ldots$ for the system. Each cell has a finite list of states, and a rule which gives the state of a cell at time $t + 1$ as a function of its own state and the states of its neighbours at time t. The list of states, together with the rule governing the state transition of a cell, is called a *transition function* (corresponding to the transformation rule of Lindenmayer).

Von Neumann devised, but never simulated, a 'self-reproducing' cellular automaton of huge complexity—each cell possessed 29 distinct states. Codd (1968) has shown that this can be reduced to a programmable 8 states per cell, and other workers have also simplified von Neumann's original model. The uses of the cellular automaton lie mainly in programming theory. For example, a growing cellular automaton that is synchronously dividing into a number of replicates might be programmed to carry out a certain computation. If the replicates can simultaneously carry out such a computation, parallel computing may be possible. Other hopes for cellular automata are that they can be used to make their own computer software—'growth' and 'repair' of automaton logic circuits might then be feasible.

This sort of approach is far from biology, and study of the problems of development which start from the basic definition of a cellular space tend to fall into two groups. In the next section we will deal with the simulations which use the ideas of arrays and self-reproduction as a starting point. Other authors have tried to view intercellular developmental processes in automata-theoretic terms. Arbib (1969) has probably done most to further the cause of the automaton in developmental biology. Arbib stresses that the problems of self-replication of interest to the automata theorist do not end with the 'largely trivial' (Arbib, 1969) DNA–RNA, and RNA–enzyme transductions. It seems much more important to find out how a complex multi-celled automaton can grow from a single cell. Arbib admits that most work at the present time is on 'ingenious programming of cellular arrays, rather than on weaving a rich mixture of theorems'. He introduces the 'Mark II Module Model'—each module is an automaton representing a cell; the biological program exists as a string of instructions stored in each 'cell', and only a portion of the control string can be read by each individual cell. The change in activation of parts of the string is the model equivalent of 'differentiation', and increase in cell number corresponds to 'growth'.

Lindenmayer (1967) looked at a simple model situation. He viewed a string of cells ('a filamentous organism') as if they were units that could undergo changes of state under inputs received from their neighbours. The cells produce outputs as determined by their state and the input they receive. Cell division occurs by inserting two new cells into the filament to replace a cell of known state and input, and all cells to the right of the divided cell then move rightwards one position. A set of 'look-up tables' was used to compute the 'next state' of each cell. For simplicity, two differentiated cell types were considered, and these may be represented by the integers 0 and 1. The most elementary example of Lindenmayer's model in action is as follows.

Growth in the filament is, as described above, from left to right, and at any time t, next states are computed by the previous state of the cell, and its immediate left-hand neighbour ('input') at time $t - 1$. The look-up table is shown below:

	Input	
Present state	0	1
0	0	1
1	11	0

If the environmental input (the left-handmost cell) is set at 0, and we start with a single 'seed' cell in the next right-hand position, the first few generations will read as follows:

Generation	Environmental input	Growth
0	0	1
1	0	11
2	0	110
3	0	1101
4	0	110111

Changing the division instruction to give (0, 1) progeny instead of the (1, 1) of this example gives rise to a repeating banded pattern as shown below:

Generation	Environmental input	Growth
0	1	01
1	1	101
2	1	0101
3	1	101101
4	1	01010101

If the environmental stimulus is also changed to 1, a 'constant basal' pattern results—an increasing string of 0's always with a 1 at the right-hand end.

Lindenmayer decided that although such unidirectional input systems may be important in isolated situations (he quotes, for example, auxin travel in plants), in many cases it is necessary to employ two directional input relationships. A further complication—the simulation of branching filaments—has also been discussed (Lindenmayer, 1967).

Applications

These simple models have proved useful in the study of the link between automata and living organisms. Simulations of pattern formation in one dimension have been carried out using the heterocyst spacing pattern in *Anabaena* (Baker and Herman, 1972; see also one-dimensional growth models, Chapter 7, and *Anabaena* growth, Chapter 3) as a model system, and in this case Lindenmayer models (called L-systems by van Dalen, 1971) have been used to explain the workings of the *Anabaena* heterocyst spacing mechanism. Lindenmayer (1971) has also used the analogy of biological processes to language theory to provide some basic ideas for a possible 'language of development', complete with axioms, strings, and sets of productions; these ideas have been closely integrated with those of the L-system.

Among the abstract models with no *formal* basis in automata theory are those of Ulam and of Eden. Eden's work has a grounding in statistics, and uses a two-dimensional square array as a spatial framework (Eden, 1960). Each point in the array is a computer storage location that can hold a number representing a cell. If the model starts with a single cell in the centre of the array, and this cell divides so that its daughter cells are allocated randomly to surrounding array locations (with the proviso that cells can only divide when they have free edges and do not divide into diagonal array positions), what sort of shape is produced, and how does direction-biased cell division affect the model? Eden found that probability theory predicts that it is *possible* for any colony shape with the same number of contiguous cells to be generated by the model. However, the probability of obtaining any particular shape decreases as the number of cells on the colony's edge (that is, cells with free surrounding array positions) increases. The result is therefore an 'average' circular colony, as seen during the growth of many bacterial colonies. If the probability of cells dividing along one axis of the array is increased, Eden showed that the axial ratio (maximum height/maximum length) of the colony is greater than the probability ratio of preferential division. If the probability ratio was 10 : 1 for horizontal : vertical growth, the axial ratio is only about one-third, much less than the one-tenth intuitively expected from the probability ratio.

Ulam was concerned more with the general problem of the complexity of patterns that could be produced from simple rules than with the probability of obtaining particular cellular configurations. Although his work is rather abstract (witness the patterns shown in Ulam, 1962, and Figure 6.3), a brief description is included here because the mechanism of generation of complex patterns is such a central problem in developmental biology. Like Eden's, Ulam's models most usually employed the subdivision of a plane surface into squares, or cells, and 'rules' for the 'birth' and 'death' of each cell were specified. As Ulam devised a vast number of similar rules governing 'growth', two such rules will suffice to illustrate the technique.

We start in the first generation with one occupied square, and define the growth rule; given a number of occupied squares in the nth generation, the squares of the $(n + 1)$th generation will be all those adjacent to only one square of the nth generation. Figure 6.3(a) shows how growth appears after five generations, using four nearest neighbours (no division occurring into diagonal array space). If the rule is extended to include inhibition by cells which touch diagonally, Figure 6.3(b) is produced after five generations.

The game of life

Even more abstract than Ulam's work is the 'game of life', devised by J. H. Conway, a Cambridge mathematician (Gardner, 1970). Again, the complexity that can arise from the recursive application of a small number of rules is shown, but in this case, the complexity is of a rather different kind, and the accent is more on the simulation of a 'living' organism than is shown in Ulam's model. Conway's game is played on a two-dimensional array, and a certain number of adjacent array positions (each position here has eight neighbours, four adjacent orthogonally, four diagonally) are designated the 'cells' of the organism. As with the models described previously in this chapter, each cell is processed at each generation according to a set of rules. In this case the rules are:

1. Every cell with two or three neighbouring cells survives for the next generation.
2. Each cell with four or more neighbours 'dies' (from overpopulation). Every cell with one or no neighbours dies from isolation.
3. Each empty cell adjacent to exactly three neighbours is 'born'.

Conway's rules can, of course, be played out on graph paper, on a draughtboard, or best of all simulated on a computer: the most rewarding way to play 'Life' is to use a computer video display screen to watch growth occurring. Some unexpected patterns result, and several of these are shown in Figure 8.1.

Two other workers have also considered models of the Ulam type, and have drawn parallels between the simple model that uses recursively

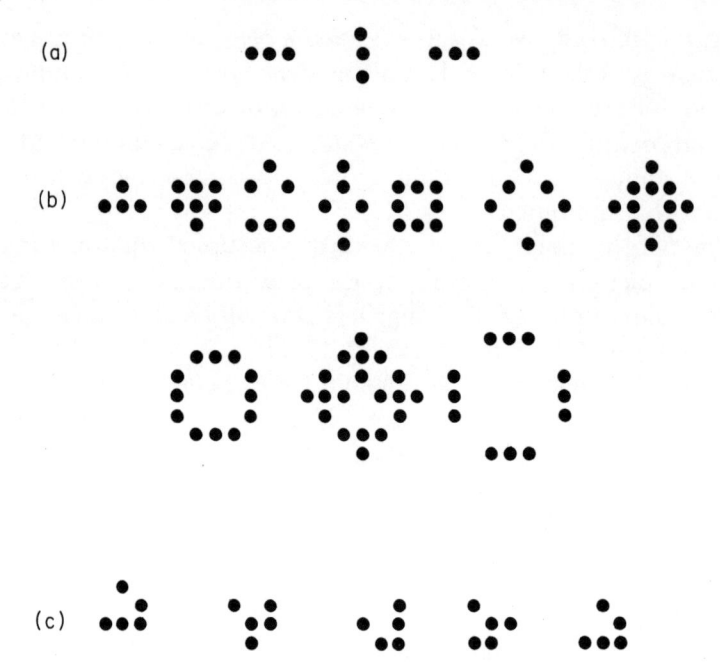

Figure 8.1 Conway's 'game of life'. The rules are explained in the text, and the three examples shown should be 'read' from left to right. (a) The pattern produced by three cells in a straight line always alternates at successive time increments as shown, and never grows. It is called a 'blinker'. (b) This pattern develops until four (a) groups are produced. These then alternate. (c) A 'glider'. After four time increments, the same pattern reoccurs one position down the right diagonal.

applied rules, and developing systems. Maruyama (1963) introduces the concept of 'deviation amplification'. Cybernetics is a science of self-regulation and equilibration but, says Maruyama, little notice has been taken of systems which generate *change* rather than homeostasis. Maruyama introduces the idea that these sorts of complex mechanisms are very relevant to development: the genes do not need to contain the whole amount of information necessary to produce the adult organism—all they need to specify are some simple rules, and interaction between the component parts of the embryo can do the rest.

'Each part of the embryo does not contain the information for all the details of what it is to become. The parts generate the information by interaction, i.e. each part receives some information, not from other parts but rather from its relationship to other parts.'

The amount of information held by the egg and its relevance to information theory* have been the subject of various analyses, for example by Apter and Wolpert (1965) and by Raven (1964).

The second Ulam-type model was developed by Gordon (1966). This model simulates the growth of a spiral generated by stochastic means. Using a two-dimensional array, Gordon defines two types of cells, I and S. Growth of these cells occurs into non-diagonal surrounding array positions. Only one S cell, called the 'leader', may divide at any one time, and the daughter cell becomes the new leader. I cells grow randomly, whilst S cells divide into the left-hand site (relative to the vector from the cell from which the dividing cell grew). If this site is not empty, then forward and right-hand sites are tested in turn and used if available. If the cell is totally trapped, then the first unsurrounded S cell nearest to the growing point divides.

Cell division occurs sequentially, and a parameter r is defined such that the probability that the next cell to divide is an I cell is rP_S, where P_S is the probability that it will be the S leader cell. All I cells have an equal probability of dividing, if they have any free sites available. Gordon shows how variation of r can give spirals with varying structures. If $r = 0.4$, for example, the spiral winds itself into a cavity. If $r = 0.1$, very wide spiral turns are exhibited, whilst $r = 0.25$ gives an equally spaced Archimedean spiral. Gordon also presents the basis for a 'general model of development', using the component parts of the simulation. The model would therefore include an 'organism' represented by an ensemble of units (cells), all capable of making decisions. For each cell, these decisions are based on the state of the cell and its environment (a

---

Information theory: developed around 1948 by Shannon and Weaver as a mathematical theory to work out the amount of variety or specificity within a system. The theory was worked out in connection with the problem of how information from a source A transmitted to a receiver C is influenced by the channel B. The most basic conclusion is that in a *closed* system in which nothing comes in from outside, yoy can never get more information in C than was initially present in A. There are several biological situations which present a parallel to the inanimate systems studied by Shannon and Weaver. Firstly, nerve conduction, where their theories have proved most useful. A little more far-fetched is the analogy of the transmission of hereditary information from one generation to the next in the genes. Here the theory is less directly applicable because of the influence of non-genetic characters on development, and because the genetic material itself is capable of change—mutation and chromosome rearrangement—which can complicate the application of information theory.

restatement of the automata-theoretic definition of a cellular space). The internal state of each cell would include its state of differentiation, a 'memory', and perhaps an internal clock.

Gordon suggests that such a general model may be used to 'evolve' computer organisms.

> 'Let an organism grow according to a set of rules R. R may be regarded as a vector whose components are a set of independent "growth constraints". Choose at random n slight variations from R, $R + \Delta R_i, i = 1, \ldots n$, and grow the corresponding organism. Those which show hints of some form or pattern which is aesthetically pleasing or biologically important are selected, and the process is repeated.'

This echoes Braverman and Schrandt (1967), who see this sort of simulation work as 'the form of a continuous dialogue between question and suggestion, arrived at through computer generations, and biological experiments and observations answering and proving these'.

CELL INTERACTION MODELS APPLIED TO SPECIFIC SYSTEMS

Cell lineage and interaction in simple organisms

Raven (1968) and Raven and Bezem (1971) provide a link between the models that have already been discussed in the earlier part of this chapter, and models more applied to specific developmental situations. Raven criticizes the approach used by the automatists—the models designed by these authors show only a remote and superficial resemblance to any real developing system. Raven suggests that this is because the automatists are not developmental biologists, and automata theory is not exactly equivalent to the processes that go on in embryogenesis. Having said this, he goes on to describe the morphogenesis of the larval head pattern in *Limnaea*, the pond snail, using terms very much like those of automata theory for the basis of this model (Raven, 1968). In this paper, he puts forward a set of simple rules for generating the snail head pattern. These rules are based on cell lineage: the internal state of each cell is dependent on that of its parent, and on the direction of the division by which it is produced. 'The cell, so to speak, "remembers" the previous history of the cell line of which it forms a part.'

Raven uses a notation to specify the developmental history of each cell, so

$$e(v, s, d_1) \rightarrow D_1$$

means that the cell in question has developed from the ovum (e), and has undergone three divisions, one dorsoventral, one sinistrodextral, and one in the animal/vegetal plane. The cell is at present undergoing another animal/vegetal division.

It is suggested that early morphogenesis in the mollusc is very dependent on the placing and direction of cleavage furrows (Raven, 1968), and that for this reason the specifications set into the model are quite valid. Raven also suggests that at this early stage, cell interactions are of little importance, and that it is the inbuilt characteristics of each cell which determine the pattern of development.

Raven and Bezem (1971) use the rules formulated above to produce a computer model of embryonic development of the *Lymnaea* egg, the model simulating growth from the 4 cell to 64 cell stage. Apart from the assumptions of the paper-and-pencil model earlier described by Raven, it was necessary to provide a geometrical base for the computer program, and this was done by representing each cell as a sphere. All cleavage planes pass through a fixed point on the egg axis, producing two daughter spheres. By varying parameters such as the angle of division, the position of the axial fixed point, and the proportion in which each cleaving cell is divided, a '64 cell embryo' was produced, and the printouts (on a graph plotter) look very similar to those previously shown in the actual analysis of the *Lymnaea* egg (Verdonk, 1968). Raven and Bezem take this to indicate that Raven's original thesis, that early morphogenesis in molluscs is determined largely from within the cells themselves, may well be true. Figure 8.2 shows a series of comparisons of the computer simulation printout with drawings from the actual developing *Lymnaea* egg (after Verdonk, 1968). As Raven and Bezem themselves conclude, it is surprising that the simulation, based only on the information inherent in each cell and devoid of any 'cell communication properties', can give such a close fit to the development of the real *Lymnaea* egg.

Honda (1973) has also used a geometrically based model, this time to show how the 'simple rules giving complex results' ideas expounded by Ulam, Maruyama, and Gordon, hold for a particular living system. Honda has analysed growing colonies of the coenobial green alga *Pediastrum biwae*, colonies of this organism being normally flat with 4, 8, 16, 32, or 64 cells per colony (Figure 8.3(a)). There are two types of cells in a colony of *Pediastrum*; horn cells situated peripherally, and triangular cells. Horn cells have three corners, one of which is horn shaped, and

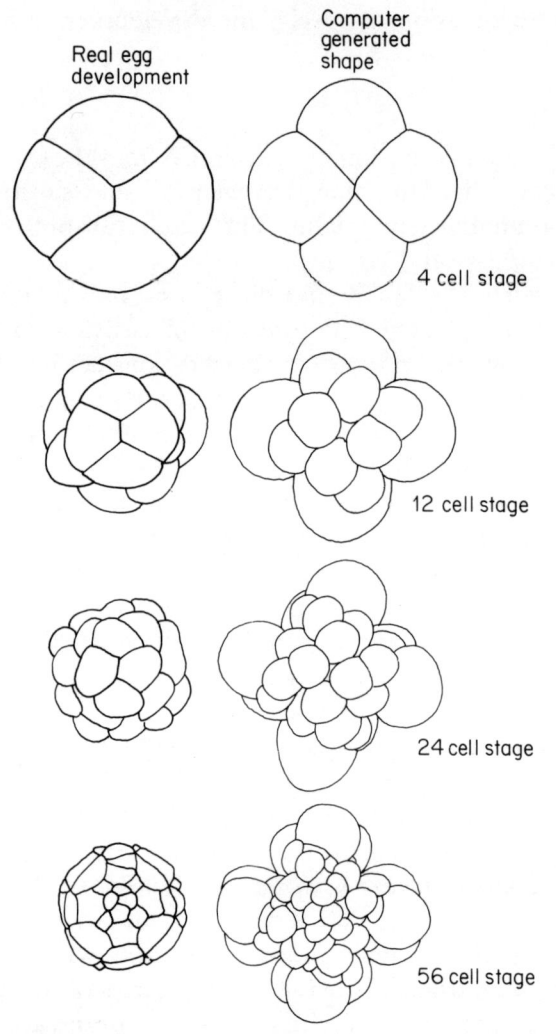

Figure 8.2 Development of the *Lymnaea* egg (after Raven and Bezem, 1971). The left-hand column shows pictures drawn from an analysis of the real egg development. The right-hand column indicates the computer equivalent at the same stage of growth.

triangular cells have three corners only. The overall number of each of these cell types does not appear to be genetically fixed: Honda suggests that the zoospores might all be of one type at first, and then become subject to a later stochastic determination.

Colony formation begins with the opening of a semicircular 'trapdoor' in each of the cells of the mature coenobium, and out of this opening come the cell contents—zoospores enclosed by a transparent vesicle. The vesicle expands, and the zoospores inside jostle around, becoming spherical in shape, as shown in Figure 8.3(b). As the vesicle expands, the zoospores aggregate into a plane, one zoospore in thickness (at this time the vesicle is almost spherical) (Figure 8.3(c)). The zoospores link to form strings and finally form several concentric circles; before mobility ceases, the zoospores change shape from spherical to conical. The vesicle finally disappears, and 'differentiation' into the mature colony is seen.

Honda suggested that the zoospores have three sites on the equators, one 'H-site', which can grow into a horn, and two 'C-sites' which cannot do so, and grow into connection points between zoospores. If there are other cells nearby which prevent the growth of the H-site, this site will grow into the third corner, and the cell will become a triangular cell; otherwise it will become a horn cell.

One of the central mechanisms in the morphogenesis of the *Pediastrum* colony is the arrangement of the zoospores into a plane. Davis (1964) suggested that the vesicle was lens-shaped, and so constrained the underlying spores. This observation has been countered by Hawkins and Leedale (1971) who saw normal colony formation within vesicles up to 3 times the width of the zoospores.

Honda has produced a computer model of zoospore alignment. The model makes two assumptions. Firstly, the H- and C-sites all have an affinity with each other. Secondly, the zoospores move so that they keep their equators at about the same angle. For example, if one zoospore collides with another, the two move together for a distance, maintaining the contact. The computer 'zoospores' move inside a simulated large spherical surface, and the initial coordinates and equatorial angles of the zoospores are determined by generating pseudo-random numbers. At each step of the program, a zoospore is chosen and it moves forward for a set distance. If it hits another zoospore, the equators become parallel to each other (this simulates the affinity of the H- and C-sites), and the new orientation direction of the equators is chosen by averaging the directional cosines of the original cells. If either of the spheres collides with a third sphere or the wall of the surrounding 'vessel' during movement, the momentum of the two zoospores ceases, and the next program step begins by random selection of another zoospore. Parameters R (radius of the containing vessel) and r (length of movement path) are varied. The problem of printing out a three-dimensional picture was overcome, as with Raven's model, by using plane projections to give a view from above and across the equator respectively (Figure 8.3(d)). With 16 cells,

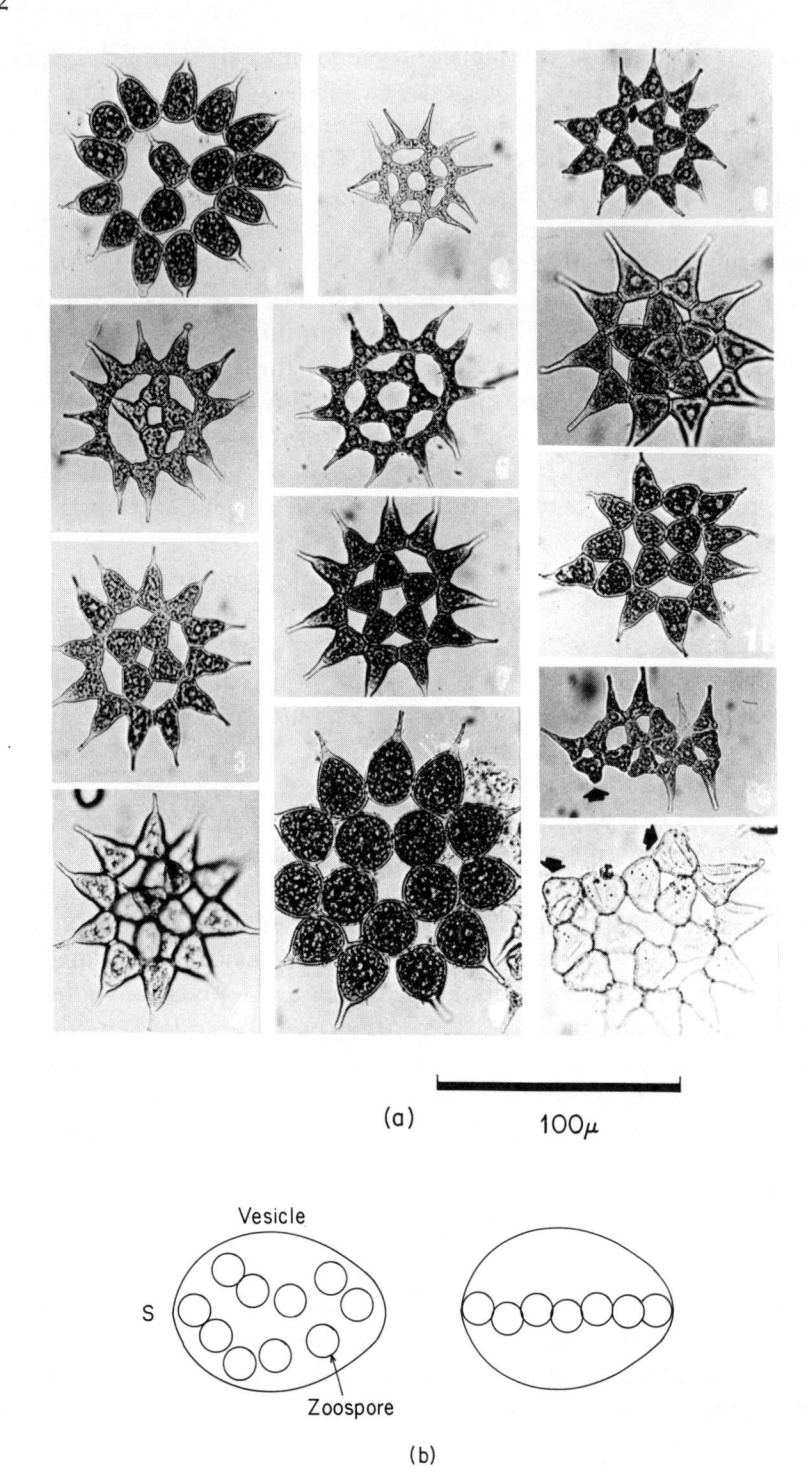

(a)

100μ

(b)

Vesicle

S

Zoospore

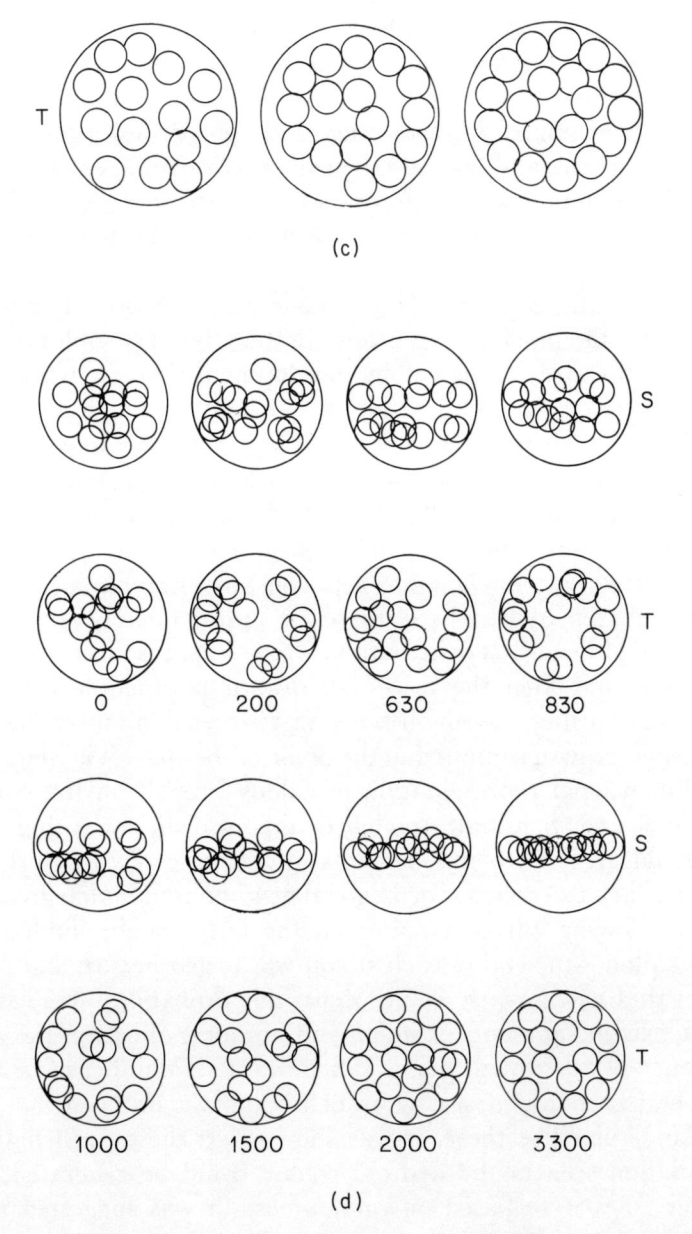

Figure 8.3 Model of zoospore alignment in *Pediastrum* (Honda, 1973). (a) View from above of various colonies of *Pediastrum*. Horn cells situated peripherally, and triangular cells positioned internally can be seen. Cells link by their three corners. (b) and (c) show formation of the colony inside a vesicle: (b) side view, (c) top view. (d) Honda's simulation results—S = side view, and T = top view. Numbers refer to number of cell movements in the simulation.

and using the rules described above, it was found that the cells arrange themselves into a flat plate after about 3300 moves—the model, therefore, indicates that the arrangement of zoospores into a plane is, like the early morphogenesis of *Lymnaea*, due to the properties of the individual units of the system, and does not depend on any overall external 'control system'. It still remains to be shown, however, that the angular buffering of the zoospore equators is actually a feature of the living organism's development.

Braverman and Schrandt (1967) have also produced a model that demonstrates the possible importance of the Ulam principles in another system, in this case during colony development of the polymorphic hydroid, *Podocoryne carnea*. This organism normally colonizes the exterior of the shell of the hermit crab, but a single hydranth removed from the shell will form a new colony on a microscope slide immersed in sea water, making it easy to see the shape of the colony which develops. The growth of the colony proceeds at a constant rate, and the ratio of hydranth number to stolon length is kept constant (Braverman, 1966). The distance between hydranths is therefore approximately constant, and follows a roughly normal distribution. Growth ensues until sexual zooids are formed, and then the rate of hydranth production slows down. Further singularities in growth are seen as the colony increases in size. As *Podocoryne* growth is limited to the plane of the slide, it is well suited to a two-dimensional representation of colony growth on the computer. Braverman and Schrandt modelled this growth by using a two-dimensional array with the elements in a square relationship to each other (i.e., each cell has four orthogonal neighbours). Stolon growth took place by allowing adjacent points on the lattice to be 'added' to the existing stolon—the end of each stolon was tested in turn and a probability set that new growth would occur: this probability was calculated from the existing amounts of stolon and numbers of hydranths within a set distance of the growth point. Whenever a set number of new stolon lengths had been added, a new branch point was initiated.

By using rules like these, it was shown that the overall patterns of differentiation seen in the hydroid colony could be generated. As the computer rules were based on randomness, it was suggested that the development of the *Podocoryne* colony also has a large stochastic component.

Cancer and models

Williams and Bjerknes (1972) have produced a model simulating abnormal clone spread from a 'tumour' through epithelial layers. The cell division activity of the epithelium is centred in the basal layer, Leblond *et al.* (1959) showing that both daughter cells of a division remain in this layer, and displace neighbouring cells. Bjerknes and Iverson (1968) have suggested that tumours are induced by an 'initiator cell' giving rise to a fast-dividing clone which takes over the entire basal layer, pushing the non-tumourous cells out of this layer. To find out how fast the tumourous cells would have to divide to do this, Williams and Bjerknes simulated the situation by using a two-dimensional array with the elements packed in a hexagonal arrangement. As the displacement of neighbouring cells of the same type would have no effect on the basal layer, it was only necessary to 'divide' cells at the tumour boundary. The computer maintained a list of boundary positions and processed cells at this boundary. Williams and Bjerknes defined a term, the 'carcinogenic advantage', which is the ratio of the speed of division of tumour cells to normal cells. If N is the total number of abnormal cells, and n is the number of abnormal cells at the tumour border, then

$$dN/dt = (k - 1)n$$

Williams and Bjerknes calculated that a tumour starting from one cell dividing with $k = 1.1$ will take 13 years to reach a size of 2 mm, on the basis of their simulations (or 1.33 years with $k = 2$).

Jorggen Carlsson (personal communication, 1972) has also experimented with computer models of tumour growth. In this case, a different question, more applied in nature, was asked. What is the effect of certain doses of radiation on tumours? Carlsson used a three-dimensional array representation of a tumour, with the irradiation parameters (for example, distance of beam, beam strength, and sensitivity of cells) built into the program. The central problem with this work seems to have been the difficulty of visualizing three-dimensional data with only line-printer output.

Models have also been constructed of cell division systems by making 'family trees' of the growing cell population (Valleron and Frindel, 1973). The model system used by these authors includes full data on the length of the various parts of the cell cycle, and can be used to simulate radioactive labelling of cell populations. Our main interest in this system is the method of maintaining information on the age and state of each cell in the population. This is done by employing a two-dimensional array $Y(J, K)$, where $Y(J, K)$ is the time elapsed between time 0 and the end of the Kth phase of the cell cycle of cell J. If g generations are simulated, the dimensions of the array will be

$$J = 2^g \quad \text{and} \quad K = 4$$

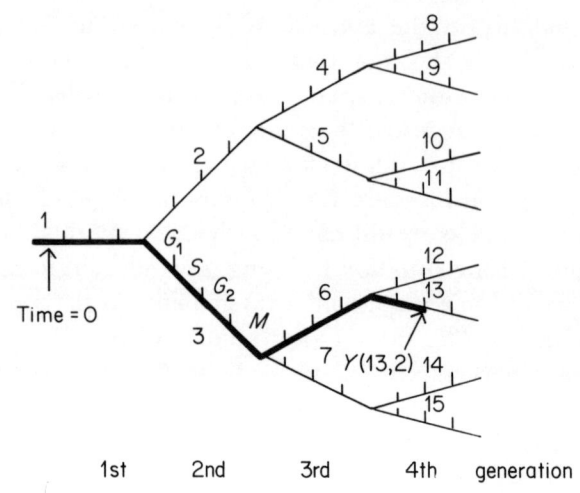

Figure 8.4 Family tree originating from cell 1 (after Valleron and Frindel, 1973). Each branch produces two new cells, and each arm is divided into the four cell cycle components. The indicated cell ($Y(13,2)$) is the 13th cell produced, at stage 2 of its life cycle.

The relationship between daughter and parent cells is that the parent cell number (J) is the integer part of $J_1/2$, where J_1 is the parent cell. The daughter cells of J_1 are therefore numbered $2J_1$ and $2J_1 + 1$. Figure 8.4 shows how this nomenclature works out in practice. K represents the cell cycle phases ($K = 1$ for $G1$, $K = 2$ for S, $K = 3$ for $G2$ and $K = 4$ for M). The times for these phases can be set as parameters.

Turnover growth is seen in the work of Düchting (1978). He constructed two-dimensional models of abstract cell division in two dimensions, and was interested in both cell renewal and in two cell types growing at different speeds. He used a 10×10 array, with a four-neighbourhood space, with the following rules:

1. If a living cell is surrounded by dead or destroyed cells it dies.
2. Cell division only occurs into an empty space next to the cell to divide.
3. If more than one empty space occurs, the division is random.
4. Cells on the array boundary only respond to their neighbours (i.e., they 'recognize' the 'edge' of the array).
5. Each cell has a pre-chosen life span.

An example of the basic model with one cell type is shown in Figure

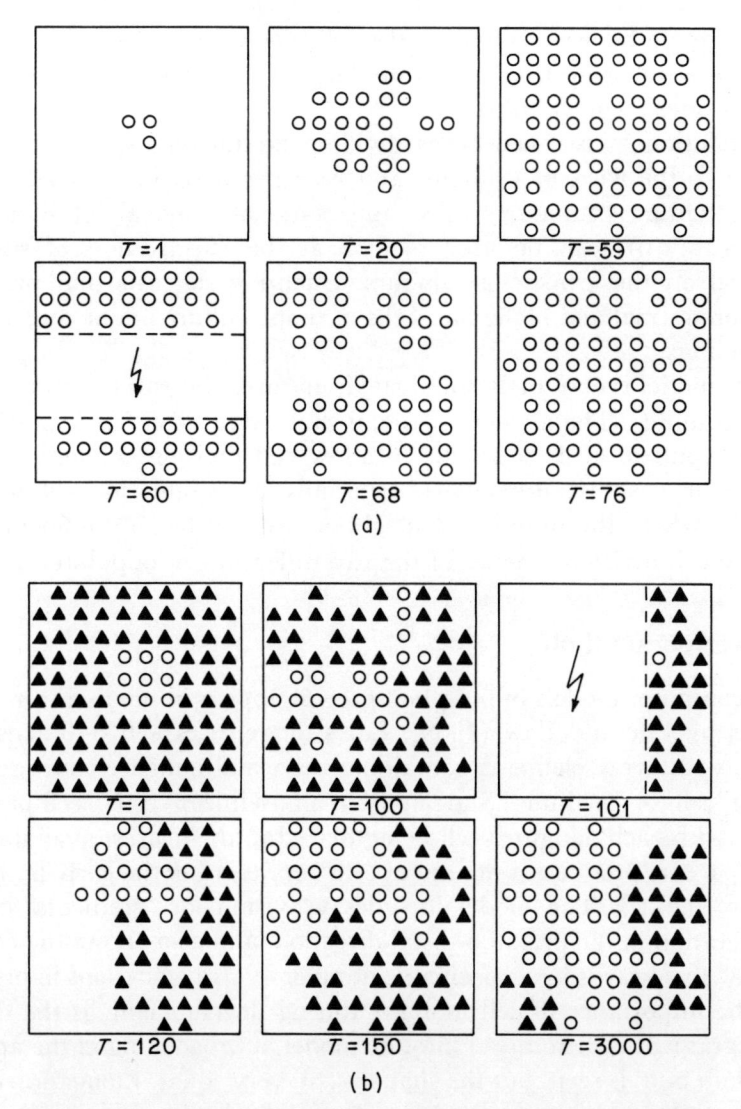

Figure 8.5 Düchting's model of cell population growth and competition in two dimensions (a) Basic model. Cells in rows 4–7 are killed after 60 time increments. (b) Simulation of two competing cell populations. The faster-growing 'cancer cells' are excised at $T = 101$ but quickly re-establish themselves by $T = 120$.

8.5(a). Three starter cells are placed in the centre of the array, and division occurs at random (cell death occurs according to the 'life' preassigned to each cell). After a number of time increments (one increment = one cell division) some cells are 'killed'. Düchting's model can therefore be used to look at the cell renewal process in his population, the idea being that this is the kind of situation that might occur after surgical removal of tissue.

Simulation of two cell types dividing at different speeds is rather similar to the work of Williams and Bjerknes described above. Use of faster-dividing ('tumour') cells, and 'surgical removal' of such cells (Figure 8.5(b)) can be used to look at the effectiveness of surgical removal on the growth of tumours. Düchting suggests that practical work on normal and malignant cells in tissue culture might corroborate his findings.

This model is extremely simplistic in nature, and can be criticized on many grounds. The array is so small that 'edge effects' (see p. 87) are over-simplified. Cells must die to allow cell division. Finally, a four-neighbour space is unrealistic, especially for simulations of tumour growth, where the number of neighbours could heavily influence the relative cell division kinetics of the two different cell populations.

The vertebrate limb

Two computer models of limb bud growth appear in the literature. The first is by Ede and Law (1969). Ede's model uses a two-dimensional array, with array elements containing the usual numeric codes representing 'cell' or 'medium'. Starting from a baseline of 'cells', cell proliferation (with each daughter cell being placed in the nearest available site to its parent) produces a mound of cells representing the early limb bud (Figure 8.6(a)). The model includes provision for 'gradients' of cell proliferation in the proximo-distal direction, and also allows the cells to 'move' distalwards by a specifiable number of array locations in order to test the importance of cell mobility during development. If the distalward gradient is introduced into the model, a broadening of the apex of the 'limb bud' is seen, but the shape is still very squat. Elongation of the stem region can be introduced by allowing slight distalward cell movement. If this movement is combined with the mitotic gradient, and the distalward movement is reduced slightly at the apex, a shape looking rather like that of the late limb bud is seen (Figure 8.6(b)). If the distalward movement is omitted, a shape looking like the *talpid* mutant is seen, as shown in Figure 8.6(c). As Ede has already demonstrated that *talpid* limb mesenchyme cells are less motile than normal cells, the model seems to fit the possible causal factor behind the mutant.

Ede's model makes no attempt to simulate the influence of cell differentiation (all cells in the model are of one type). Indeed, the division

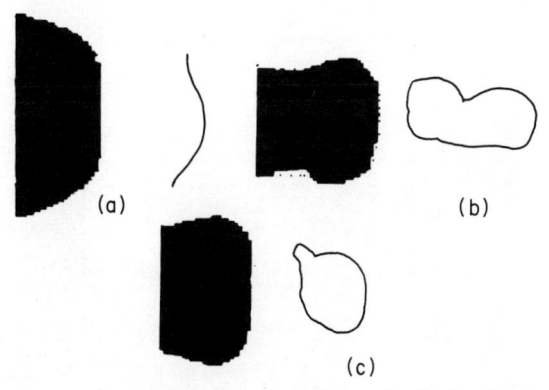

Figure 8.6 Computer model of limb bud development (after Ede and Law, 1969). Black shapes are computer generated limbs, and outlines represent actual limb bud shapes. (a) Simple cell proliferation in the computer model gives an approximation to the early stages of limb bud formation. (b) Distalward cell movement + mitotic gradient in the computer model gives the 'paddle' shape of later limb bud growth. (c) No distalward movement in the computer model results in a shape approximating that of the *talpid* mutant.

algorithm used would only work with one cell type—any form of clonal growth requires 'space clearing' next to the parent cell (see Chapter 7, p. 94). The model also restricts itself to two dimensions. Flow charts for the model are shown in Figure 8.7.

In programming terms, Ede's model is very basic, but deserves further consideration because it was the earliest practical application of modelling techniques to a complex developing system. The array is square in nature, and the 'space clearing' problems are overcome by placing the second daughter cell in the nearest available free array location to a 'search focus', a set number of rows forward from the cell to divide. Apart from the unrealistic separation of the two division products, Ede and Law seem to have scanned the array for cells to divide in a sequential manner, introducing bias into the system. The rule for daughter-cell placement does not suffer from this shortcoming, as an area around the search focus may be scanned for a free site in four possible directions, the direction being changed at each cell division.

The second model system of limb bud growth, by Mitolo, also suffers from the disadvantages of having no 'space clearing' algorithm. In this

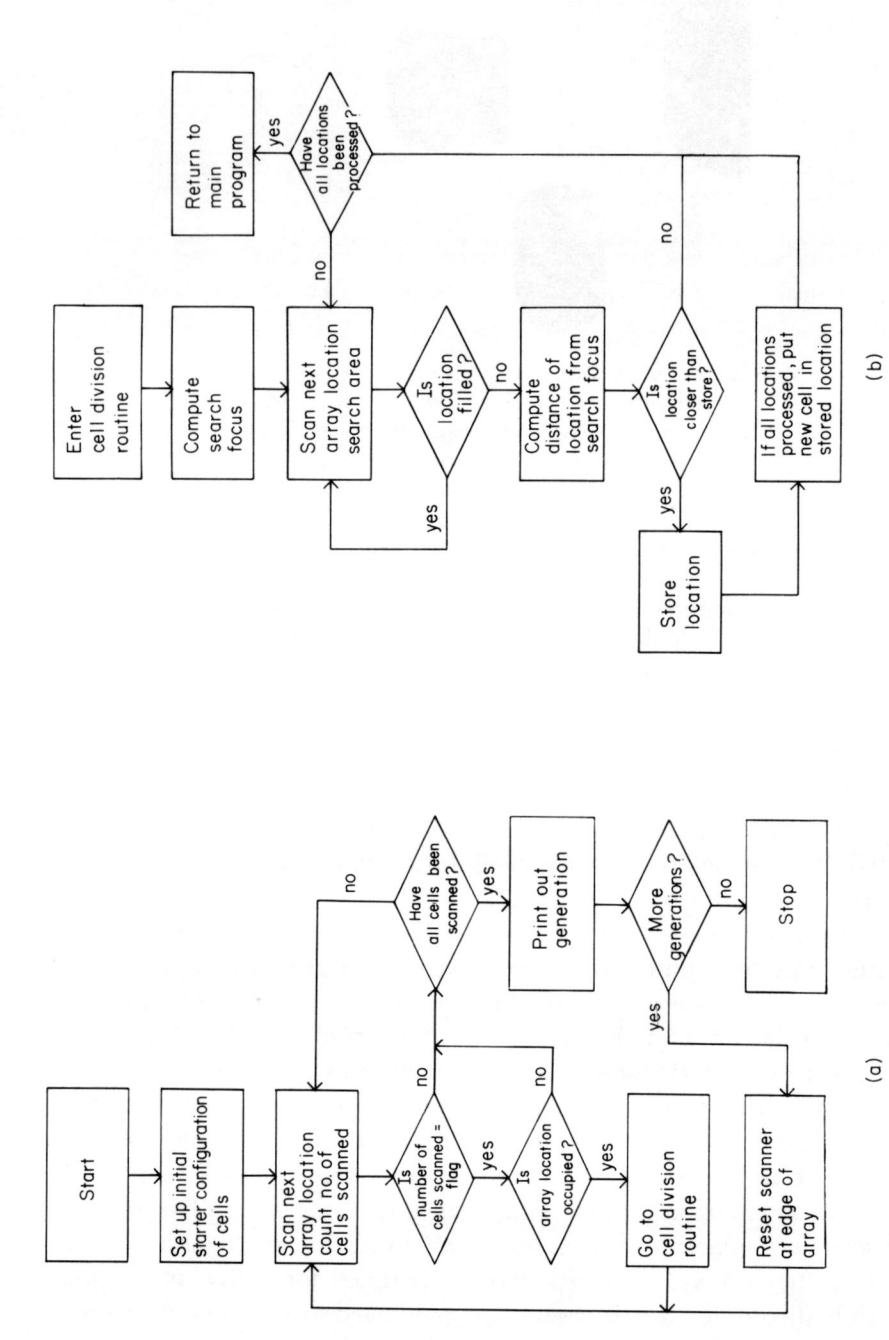

Figure 8.7 (a) Flow chart for main Ede and Law program. (b) Flow chart for cell division routine.

case, however, a slightly different form of simulation is used. Mitolo's model (Mitolo, 1971a, 1971b) also uses a one-cell type system, and places great emphasis on the relative angles to the 'baseline' at which the cells of the growing limb bud divide. An array format is also used to give a spatial coordinate system. An axial direction of growth from the baseline is chosen, together with a standard angular deviation from this direction: this deviation governs the variation from the mean division angle that any one cell can make when dividing.

Mitolo also includes a discrete proximo-distal gradient of cell division in the model, and concludes that a gradient roughly in the proportion of 1 : 2 with two-fold more divisions in the distal region is necessary for realistic simulation. In addition, the growth angle should be set at 60° to the base and a standard angular deviation of cell divisions of 40° is also used to produce an accurate-looking two-dimensional limb bud (Figure 8.8).

In a later paper (Mitolo, 1973), this simple model is taken further. Mitolo has found that his own model, like Ede's, could simulate development of the *talpid* mutant. In this case, the rule was simply that cells should divide completely at random, and possess no specific division direction. This appears at first sight contrary to the Ede model, but not so: distalward cell movement was programmed in the latter case by making the dividing cells place their offspring in the nearest available *distalward* array position, and this is equivalent to division directly away from the baseline in Mitolo's model. Mitolo's normal limb buds show

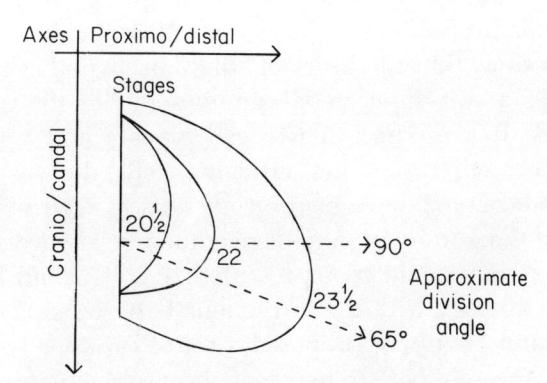

Figure 8.8 Mitolo's computer-generated limb bud patterns. Three stages of the equivalent 'real' limb bud growth are shown. A division angle of 65° to the baseline was used.

similar growth to Ede's, except for the fact that the angle between baseline and outgrowth direction is 60°, and the broadening of the bud apex is missing. Both models therefore lead to a similar conclusion: that absence of distalward cell movement (Ede) or of defined angle of growth (Mitolo) result in the *talpid* mutant.

Practical experiments on the living system seem to bear out these results. If normal embryonic limb cells are separated and cultured on a plastic dish, they move around before settling down to form a fused network. In this case the cells are elongated with a ruffled membrane at the leading edge. *Talpid* cells are different. They are flatter, with cytoplasmic extensions all around the cell. These extensions attach to the substratum in culture, and prevent the cells moving around. It has been shown that *talpid* and normal cells in the developing limb itself also show these same differences.

These computer models show two things. Firstly, even crude models can simulate real systems accurately enough to suggest ways in which particular developmental processes operate. In addition, two different ways of approaching the same problem can result in rather different descriptions of the same formal process. Ede's simulation differs from Mitolo's but formally shows basically the same thing. There may be many model representations of a system, but they may all be interpreted in terms of one and the same mechanism.

The aggregation of cells

Cell aggregation has been studied by several groups of workers. It was initially Goel and his co-workers (Goel *et al.*, 1970; Leith and Goel, 1971) who set out to show that the 'work of adhesion' hypothesis of Steinberg (1963) held for a two-dimensional simulation. Steinberg's hypothesis was as follows. If a mixture of two cell types a and b is allowed to re-aggregate after separation into individual cells, the way in which the re-aggregation occurs is dependent on the *work of adhesion* possessed by each of the cell types in relation to its neighbours. The work of adhesion for a two-cell system could be represented W_{aa}, W_{ab}, and W_{bb}. Work of adhesion is an analogy with the attractions between molecules in thermodynamics, and Steinberg proposed a set of rules for re-aggregation, dependent on the relative strengths of the attractant forces: if $W_{ab} > (W_{bb} + W_{aa})/2$ then there will be an intermixing of the two cell types. If the opposite, $W_{ab} < (W_{bb} + W_{aa})/2$ is true, then segregation will occur. If $W_{bb} > W_{ab}$, then complete sorting out into two separate populations will take place. If $W_{aa} \geq W_{bb}$, then what Leith and Goel (1971) call an 'onion'

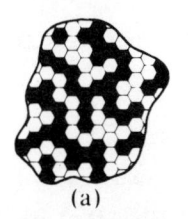

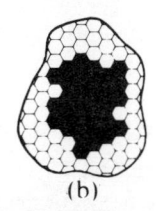

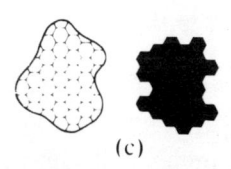

Figure 8.9 Steinberg's 'work of adhesion' model. Black cells are a-type. The patterns of cell adhesion shown are produced by the following conditions.

(a) $W_{ab} > \dfrac{W_{bb} + W_{aa}}{2}$, (b) $W_{aa} \geqslant W_{bb}$,

(c) $W_{ab} < \dfrac{W_{bb} + W_{aa}}{2}$

configuration will occur, with the a cells forming a ball inside a surrounding spherical mass of b cells. (A pictorial representation of these rules is shown in Figure 8.9.) This occurs because the system will tend towards a minimum surface area, and a minimum free energy per unit of this area. If the thermodynamic reasoning is correct, then only having all the cells of type b on the free periphery of the sphere will give the minimum configuration.

This hypothesis is attractive, and neatly fits Steinberg's (1963) experimental results. However, an alternative mechanism has been proposed by Curtis (1962), who suggested that re-aggregation is a kind of 'repair' system: when cells are disaggregated, they lose their surface attractions, and re-aggregation is merely a function of the time at which cells reacquire their 'stickiness'. Cells recovering quickly would form the 'central' cell type in an aggregate, whilst cells recovering more slowly would aggregate peripherally.

There is also a third mechanism proposed to explain re-aggregation phenomena, due to Townes and Holtfreter (1955). Whereas Steinberg and Curtis both invoke phenomena dependent solely on cellular interactions, Holtfreter suggested that specific chemical messengers are involved. Cells attracted to a particular chemical signal might form the 'central' cell type in an aggregate.

Goel et al. (1970) devised a two-dimensional model simulating cell aggregation processes. Their model uses a four-neighbour cellular space, and each contact edge between two cells is assigned a binding affinity, λ. λ is equivalent to W in Steinberg's terminology; thus λ_{aa}, λ_{ab}, λ_{bb} may be assigned. An 'E function', roughly analogous to the surface free energy of Steinberg, is defined as the number of edges of each computer cell multiplied by the cells λ-value in a given pattern. The E function for a pattern is therefore

$$E = \lambda_{bb}N_{bb} + \lambda_{aa}N_{aa} + \lambda_{ab}N_{ab}$$

Using this terminology, together with 'rules' for movement of cells, a two-dimensional equivalent of a realization of Steinberg's hypothesis may be tested. From Steinberg's work, if

$$\lambda_{bb} < \lambda_{ab} < \tfrac{1}{2}(\lambda_{aa} + \lambda_{bb}) < \lambda_{aa}$$

then the only stable pattern will be when N_{aa} and N_{bb} are at a minimum, and this will only occur if the a cells are packed inside the b cells: this is the 'onion' pattern. The validity of the model can therefore be tested to see if 'onions' can be produced. In Goel's model, λ_{aa} was made to equal 1, and the other affinities equalled zero: a 20×20 lattice was used, and border effects (see Chapter 7) were avoided by placing a two cell width 'no-man's-land' around the lattice. The a cells are all scanned once during each discrete time step, and each cell tries to maximize N_{aa}. This is done by calculating the increase ΔN_{aa} that would occur by movement to any of the neighbouring locations. If N_{aa} can be increased, swapping of the picked cell and the cell in the chosen array position occurs. Goel's first paper (Goel et al., 1970) concludes that the simple application of rules of this type is unable to give the 'onion' pattern. The situation appeared to be rescued by a second paper (Leith and Goel, 1971) in which onion configurations *were* obtained. Leith and Goel investigated a large number of different motility rules, and concluded that the production of 'onions' depends on the ability of cells to 'feel' the presence of other cells several cell diameters away. Not only this—the cells in the new model could also move over two or three cell diameters by 'hopping', and subsequent swapping with the cell in the required site.

These rules have been criticized by Antonelli, Rogers, and Willard (1973). Antonelli, in fact, throws doubt on the validity of simulating the onion pattern at all, stating that Elton and Tickel (1971) have shown that central clumps in real cell systems may, in fact, be more dispersed. Antonelli's group have produced their own model (Antonelli, Rogers and Willard, 1973) and this uses a hexagonal array, which seems to be a more realistic format for simulating cell patterns in two dimensions. Antonelli quotes Fejes Toth (1964), who has constructed a theorem stating that convex planar 'cells' in maximal contact are circular in shape, have a fixed radius, and are arranged in a hexagonal 'beehive' pattern. Antonelli's model is founded on the *exchange principle*, which says that any neighbouring cells may change position by 'swapping' only if

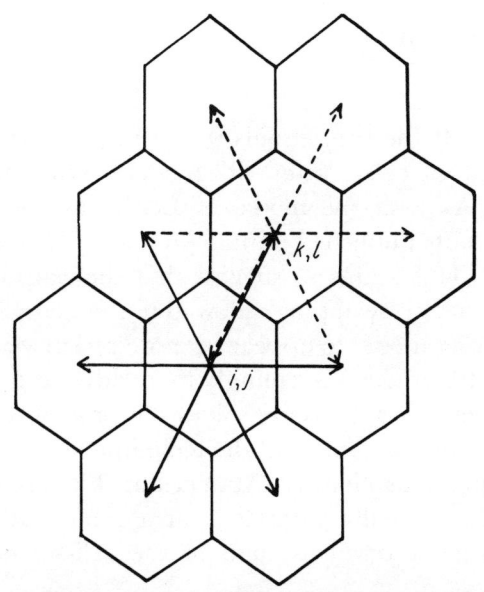

Figure 8.10 Antonelli's i,j rosette system (Antonelli, Rogers, and Willard, 1973). The cell with coordinates i, j is chosen, and it inspects its six neighbours. On picking a neighbour (with coordinates k, l) the neighbour immediately signals whether or not it is of the same type as the cell to move or not. If it is not, it adds up the number of cells of its own type in the surrounding six sites, and the number of cells of its own type that it would have if it moved to the i, j position. The change $\Delta_{kl} = kl_{(\text{after})} - kl_{(\text{before})}$ is then computed.

the interchange is beneficial to both. Finding out whether the move will be beneficial to both is done by what Antonelli terms the 'i, j rosette system' (see Figure 8.10). The cell with coordinates i, j is chosen, and it 'inspects' its six neighbours—on picking a neighbour (with coordinates k, l) the neighbour immediately signals whether it is of the same type as the cell to move. If it is not, it adds up the number of cells of its own type in the surrounding six sites, and the number of cells of its own type that it would have if it moved to the i, j position. The change,

$$\Delta_{kl} = kl_{\text{after}} = kl_{\text{before}}$$

is then computed, and the same is done for the i, j cell. All six neighbouring sites of the i, j cell are processed in this manner, and if none of the sites fulfil the requirement,

$$\Delta_{ij} + \Delta_{kl} = 0$$

no move is made. If the sum equals 0, the move is 'random', with no benefit to either of the cells. Antonelli's model is simulated on a 40 × 40 hexagonal grid. As with the model of Leith and Goel, $\lambda_{aa} = 1$, and $\lambda_{bb} = 0$. In the results published (Antonelli *et al.*, 1973), no large central clumps are seen. This seems to suggest that the central clumping is a feature of the extended 'swapping' allowed in the Leith and Goel model.

The most exciting model to appear in more recent years (Matela and Fletterick, 1979; 1980) uses a seemingly more advanced approach to the cell-sorting problem. It also partly solves the problems associated with rigid array bound simulations, and the technique used has already been outlined in the previous chapter. Matela and Fletterick did not use a rectangular or hexagonally tesselated array, but rather used graph theory to construct a trivalent map of the cell-sorting system. The

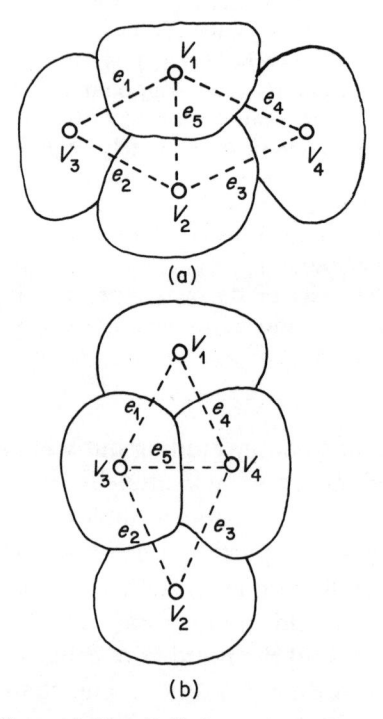

Figure 8.11 Cellular representation of the exchange mechanism from the Matela and Fletterick model (1979).

advantage of this approach is that cells may change neighbours without having to hop over several cell diameters to effect an exchange. Although this at first sight seems similar to the Antonelli model, it also has the additional advantages that the cellular matrix is non-rigid—cells may have both more or less than six neighbours—and the cells are also deformable. The cells in the model are represented by a geometric map of closely packed n-gons as shown in Figure 7.17. Cell exchanges are carried out in the following manner. Consider a system of four cells (Figure 8.11 (a)). The edge of e_5 is at first between two cells V_1 and V_2. What will happen if contact is lost between cells V_1 and V_2 and cells V_3 and V_4 make contact? Figure 8.11(b) shows the result. The edge $e_5 = (V_1, V_2)$ then becomes $e_5 = (V_3, V_4)$. One bond has simply been replaced with another. By using a series of abstract points, a plausible method of modelling changes in cell contacts has been achieved.

In simulations of cellular rearrangements, two types of cells (B and W) are used, one of which shows preferential adhesions to cells of its own type. Adhesions between B–B, B–W and W–W cell pairs are controlled by means of binary look-up tables. A series of look-up tables are available, and the particular one chosen depends on comparison of two values ϕ_{12} and ϕ_{34}, which are calculated as the number of B cells adjacent to the four-cell 'quadrilaterial' in Figure 8.11. A series of alternative expressions are therefore produced, depending on the neighbouring cells:

$\phi_{12} = \phi_{34}$
$\phi_{12} < \phi_{34}$
$\phi_{12} > \phi_{34}$

and the choice of look-up table is made on this basis. The structure of the look-up tables is as follows:

		New edge type		
		B–B	B–W	W–W
Existing edge type	B–B	1	0	0
	B–W	1	0	1
	W–W	1	0	1

Exchanges will only occur where the table entry is a 1, so if an existing edge has a B–W bond, this would only change to B–B or W–W.

Another advantage of this type of model is that the 'rules' for cell adhesion are very simple. Unlike the complex 'edge functions' of previous models, simple binary choices are made between cell states, and this type of system could correspond to an equally simple biological control system on the cell surface. Although the Matela model does seem to simulate cell sorting of two cell types in two dimensions accurately (Figure 8.12), its main usefulness may be in other fields of developmental simulation, as discussed in Chapter 7.

Gordon *et al.* (1972) have investigated a model which invokes the idea that cell sorting is carried out by the tension at interfaces between cells, and is resisted by the presence of 'tissue viscosities'. An analogy is drawn between this process and the kinetics of the break-up of an unstable emulsion between two immiscible liquids. The kinetics of cell aggregation have also been studied by a Russian group led by A. V. Vasiliev (Vasiliev, Pyatetskiy-Shapiro, and Radvogin, 1975). This simulation is

(a)

(b)

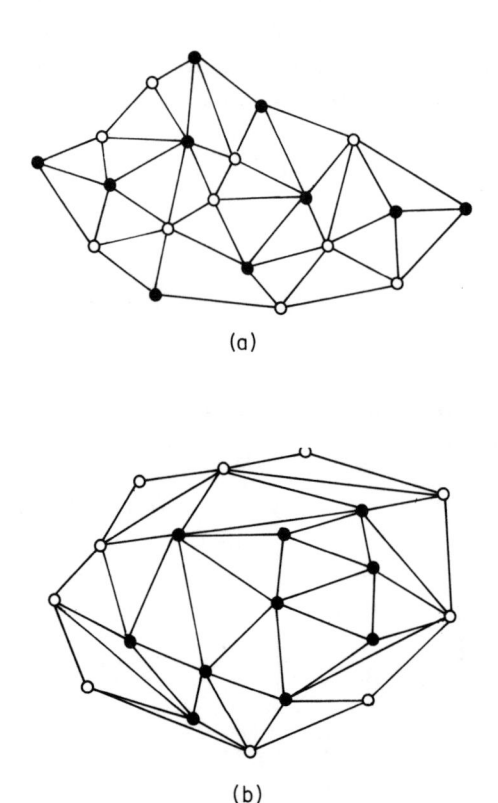

Figure 8.12 Cell sorting with the Matela and Fletterick model.

very similar to that of Leith and Goel, and also invokes the long-range forces criticized in the latter model.

To summarize the various conclusions of these authors, it seems that the 'onions' initially hypothesized by Steinberg may be due to some other factor than simple differential stickiness of two cell types. Antonelli's group, using local exchange rules only, were unable to produce concentric aggregates. Antonelli thinks that the geometric configurations taken up by aggregating cell masses are not as easy to understand as Steinberg led us to believe in his original (1963) work, and that some non-random starting effects may be necessary to produce 'onions'. On the other hand, Matela has shown that a large degree of isotypic cell association may be produced by using a topological model, highlighting the importance of the correct choice of computer program. The problem also arises as to the validity of simulating two-dimensional cell aggregation at all. Whereas with most three- to two-dimensional compressions it can be argued that only certain aspects of the system are to be simulated, the work-of-adhesion hypothesis, with its 'thermodynamic' overtones, may well not hold in two dimensions even if it works in three. Goel's group did a few simulations on 'square' three-dimensional arrays, but the small size of these ($10 \times 10 \times 10$), with the tremendous edge effects associated, may invalidate the finding that three dimensions behave essentially as two.

MODELS OF PATTERN SPECIFICATION

The models described above have all consisted of masses of simulated cells which interact together—they move and divide. Often, however, it is required to model not the movement of cells, but the way in which a pattern-specifying signal can be distributed over an initially uniform field of cells. Although the resulting pattern will be exhibited as cell-specific differentiation in the real system, pattern-specification models often consider cells as static entities whose mobility is of little importance for simulation purposes.

There are a number of computer models which fit into this category. Waddington and Cowe (1969) have simulated the formation of the pigmentation pattern on the shell of the mollusc *Olivia porphyria* ('tent olive'). The shell pattern is shown in Figure 8.13(a). Growth of the shell occurs by constant production of shell material at the growth edge, and the pigmentation is laid down simultaneously with the shell at this edge. As can be seen, the pigment is produced in continuous lines, and

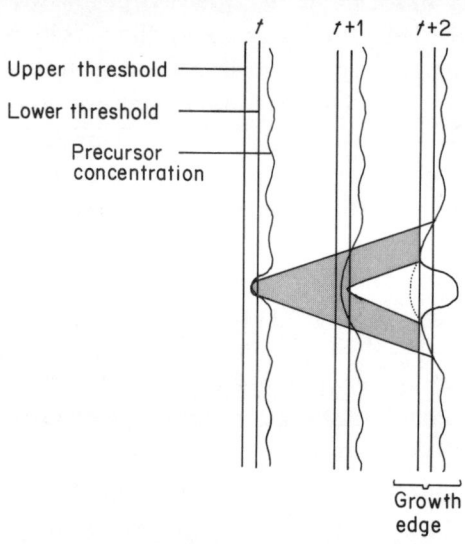

Upper threshold
Lower threshold
Precursor concentration

t $\quad t+1 \quad t+2$

Growth edge

(b)

(c)   (d)

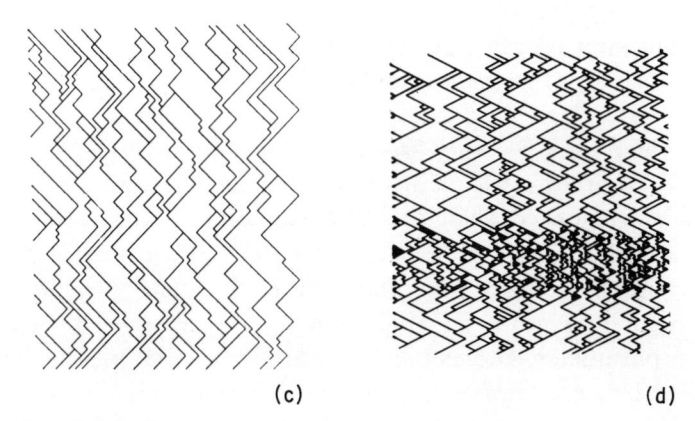

Figure 8.13 Simulation of the pattern on the shell of *Olivia porphyria* (a). (b) shows the mechanism proposed. (c) and (d) show the results of running the computer program. (Shell kindly lent by the Royal Scottish Museum.)

Waddington and Cowe have suggested that pigment initiation points arise stochastically, and from each initiation point two lines of pigment diverge. These lines gradually get further away from the growth edge as it lays down new material, giving a pattern looking like a network of 'tent-like' shapes. When two lines of pigment intersect, pigment deposition ceases. Waddington and Cowe suggested that the mechanism explaining this pattern formation process is based on the relative concentration of pigment precursors in the cells at the growing edge of the shell. The concentration of pigment precursor is hypothesized as being randomly distributed throughout the growth edge. If the precursor concentration rises above a certain value at a point, pigment deposition starts. At a later increment in time, lateral diffusion of the precursor broadens the area of pigment deposition. Another essential feature of the model is an upper threshold above which the precursor is destroyed or reduced in level; if this concentration is reached in the centre of the high-concentration region which is laying down pigment, deposition will stop at this region, and two diverging lines will be formed (Figure 8.13(b)). If two of these lines meet, the addition of their precursor concentrations carries both above the upper threshold, and deposition ceases in both lines.

The computer simulation of this model uses a video display screen to enable the operator to watch the manner in which the pattern is built up (see Appendix 1 for details of video techniques). The use of video enables interaction between user and computer, so that the operator varies the parameters until a realistic looking pattern is produced.

The computer program simulating shell pigment patterns is based on an array format, as were most of the cellular models described in the last section. Particularly important is the right-hand-most column of the array, which act as the 'growth edge' depositing 'lines of pigment'—these lines subsequently move leftwards across the array as the column of 'deposited shell material' advances across the array. The irregular variations in precursor concentration are simulated using a pseudo-random number generator—for each initiation point, two lines of 'pigment' diverge at an angle which can be varied by changing a program parameter. It was found that the first attempts at the simulation failed because large white triangular areas, a characteristic feature of the shell pattern, were missing (Figure 8.13(c)). The angles on the pigment lines of the actual shell also pointed in a 'growth direction', i.e. the angle of each joining pair of lines was canted over to the left or right, the whole pattern 'leaning' in the same direction. The program was amended so that if new initiation points arose within a certain distance of an existing line, the point was to be shifted on to this line. The pattern produced using this rule looked very similar to that seen on the actual shell, as shown in Figure 8.13(d).

Waddington and Cowe point out that this pattern reproduction indicates only the *logical structure* and not the *causal mechanism* behind the production of the shell pattern. Their work has been carried further by Herman and Liu (1973), who have tried a severe test of the model—does it work for other related but different-looking shell patterns? Herman and Liu used a version of the Lindenmayer model described on page 13, with the most recently generated Lindenmayer filament representing the 'growing' edge of the shell (the right-hand-most column of Waddington and Cowe's simulation). Rows of the Lindenmayer filament that are generated in successive time intervals represent the whole shell, and by changing a simple 'input parameter' two other shell patterns are produced. One of the patterns shows large triangular dark areas, where pigment is deposited over a wide area for a period. The flat bases of the triangles means that cessation of pigment formation is very sudden, and is synchronized in some way. Herman and Liu draw an analogy between this situation and the 'firing-squad synchronization' problem in automata theory. This problem concerns how a finite-state automaton may be programmed in such a way that each element of a one-dimensional array will enter a specific state simultaneously, irrespective of the length of the array (Herman and Liu, 1973).

Herman and Liu have also simulated the 'French flag' model by using Lindenmayer automata. The 'French flag' model was first proposed by

Wolpert in 1969 to show how pattern specification might work in perturbed systems, and its workings have already been investigated in Chapter 3. The simulation of the 'French flag' seems to be an example of the 'wrong' sort of simulation to do (or, at any rate, the wrong sort of simulation to publish). As an intellectual exercise, devoid of any insight into developmental biology, it is relatively easy to devise methods by which regulation in the flag could be accomplished (see, for example, Apter, 1966). From a brief acquaintance with the problem, one could see that an explanation requires a signal and a response mechanism of some sort, and any further insight into the actual workings of an *ad hoc* mechanism can hardly be expected to help sort out how regulation works, unless of course it is more specific and uses parameters demonstrated, or expected to be present in the real system.

Herman and Liu's simulation works using reflecting waves of three different types. When one of the end 'cells' of a one-dimensional array is disturbed, it sends out waves i, j, and k with relative speeds 1/1, 1/2, and 1/3 respectively. When the i-wave propagates to the opposite end and reflects back, it meets the j-wave and the k-wave and the boundaries between the various parts of the flag are set up. If the array is cut, a new series of i, j, and k waves are produced to build a new French flag.

Lawrence and co-workers (described in Lawrence, 1971) have directly simulated a system which exhibits gradient-like properties, by using the cuticle of the bug *Rhodnius* as a marker for the activity of the underlying secretory epidermal cells. The polarity of the individual epidermal cells is expressed in the oriented cuticle secreted by the cells (Figure 8.14(a)). Locke showed in the late 1950s (see Lawrence, 1970) that the laterally oriented ripples seen on the cuticle seemed to be controlled by a gradient. If a square piece of epidermis from a larva in the final larval stage is rotated through 90°, the epidermis secretes an S-shaped curve of ripples in the adult cuticle (Figure 8.14(b)). The surgical operation involved is shown in Figure 8.14(c). The configuration of the S depended on whether or not the rotation was clockwise or anticlockwise. Lawrence (1971) proposed that this result was caused by the presence of a concentration gradient of a 'morphogen' (Turing, 1952; see also Chapter 3). If cells from a region of high concentration are placed near to cells containing a low concentration, diffusion occurs between them, and an intermediate concentration value is formed.

A computer model was devised to simulate such a concentration gradient, and to see if rotation through 90° would give a contour map like that seen in the *Rhodnius* cuticle. Several variations of the model were tried, to answer questions about how the supposed 'gradients' might be

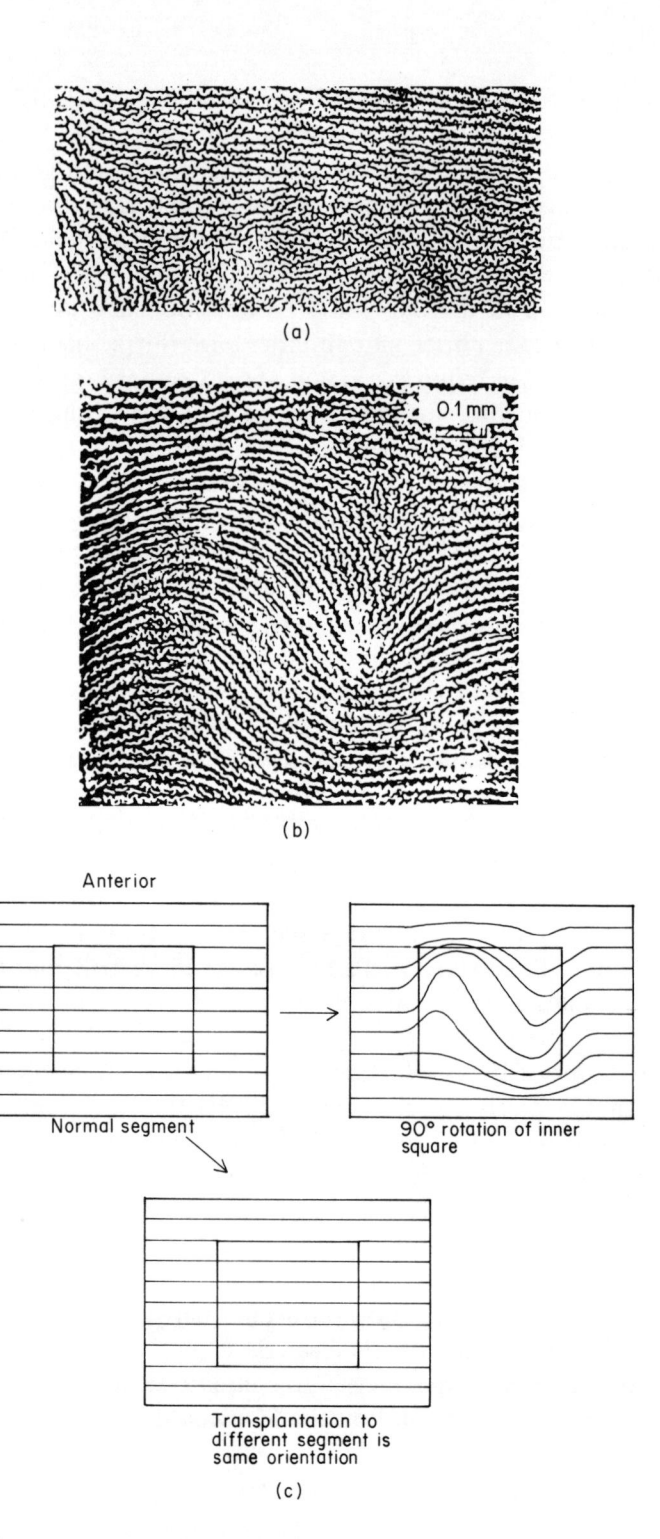

(a)

0.1 mm

(b)

Anterior

Normal segment

90° rotation of inner square

Transplantation to different segment is same orientation

(c)

(d)

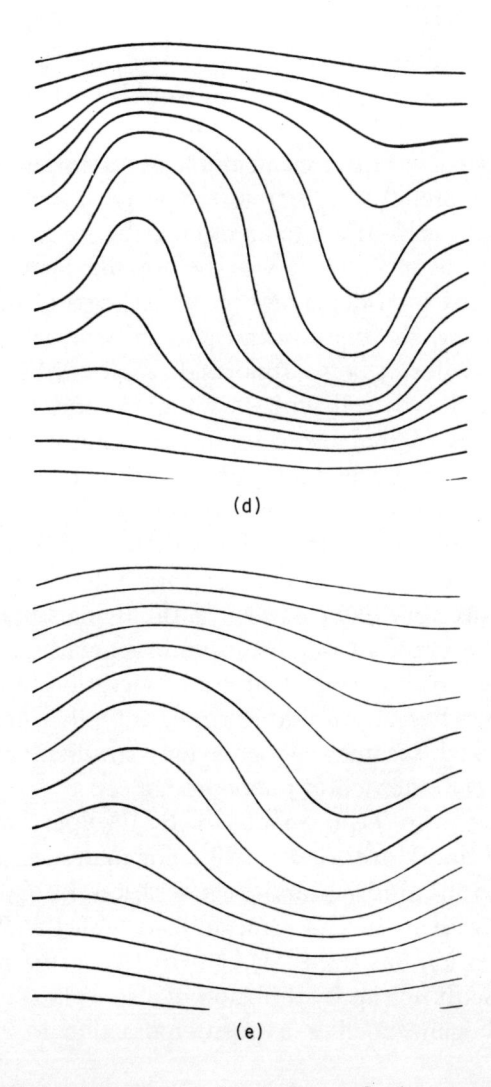

(e)

Figure 8.14 Computer simulation of the rotation of squares of *Rhodnius* cuticle. (a) Cuticle from an adult abdominal tergite of *Rhodnius* showing the oriented ripples. (b) Result of rotating a square of cuticle by 90°. (c) Surgical operation involved in the rotation. (d) Computer simulation of the rotation using diffusion model. (e) Final equilibrium patterns using homeostat regulator.

acting. Firstly, does the gradient depend only on the action of cells at the margins of the segment? The results of the model designed to test this showed that S-shaped patterns similar to those seen in the cuticle were formed (Figure 8.14(d)). The model also predicted that the greater the period between rotation and cuticle secretion, the more normal should be the cuticle (i.e., the diffusion gradient would slowly work its way back to normal). This was not found to be so in the experimental situation. It was decided that some equilibrium must be maintaining the gradient landscape. A new model was therefore devised in which a 'homeostat mechanism' in the cells tried to maintain the concentration of morphogen at the same level as it was before the operation. The final equilibrium reached then depended on the success of this resistance: if the resistance is low, the equilibrium pattern will be S-shaped; if it is high, the pattern will not be so abnormal (Figure 8.14(e)).

The simulation fitted well with this theory, and Lawrence's group then further suggested that the cells try to maintain a 'level' set at a specific stage of the cell cycle, and that the cell maintains this level until it again moves through the cell cycle. If, during this time a new level is set, the acquired new level in the cell will then be different from the old. As with Waddington's model, Lawrence shows that a *logical* mechanism can give a structure very like that seen in the living situation. However, the support of the gradient dogma in embryogenesis may lead to the obscuring of other mechanisms. Careful observation of *Rhodnius* cuticle grafts after rotation has shown that the graft actually rotates itself. Thus, what initially looked like good evidence for a gradient system turns out to be a complex cell interaction phenomenon.

Gradients have also been modelled by Wolpert and co-workers (Wolpert, Clarke and Hornbruch, 1972). For many years, the results of the experiments on grafting and regeneration in *Hydra* have been explained by a gradient mechanism, and, as with the *Rhodnius* cuticle, positional information has been invoked. In *Hydra*, the positional information is supposedly set up by diffusion of a morphogen from the head end. The morphogen specifies a positional value to which the cells respond.

The *Hydra* may be arbitrarily subdivided into nine regions *H 1 2 3 4 B 5 6 F*, with *H* representing the head, *B* the budding region, and *F* the foot. If, say, a segment containing *4 B 5 6 F* is removed, and another head is grafted on to the cut end, and the first head is removed to give *1 2 3 / H* at various times, when, and under what conditions, will a new head regenerate from the *1* region? It is shown that if the second head is grafted on 8 hours before the removal of the host head, new host-head growth is inhibited, but if it is grafted on after removal, the host head re-forms. If only a *1 2 / H* is used, however, grafting can be carried out up to removal of the host head, in order to inhibit host-head re-formation. To find out if these findings were compatible with the diffusion-gradient theory, a computer simulation was used. The computer model used a source–sink mechanism to provide a linear gradient in morphogen concentration, and tested the outcome of grafting experiments with the theory. A line of 'cells' representing a hypothetical hydra 20 cells in length was set up with all 9 regions consisting of 2 'cells', excepting the *1* region, which contained 4 cells (measurements of the actual length ratios of each *Hydra* section were used to determine these values). The Crank–Nicolson procedure (Crank and Nicolson, 1947) was used for solving the diffusion equations. The concentrations of morphogen at the head end was set at 100 arbitrary units (this was the 'source' end), and the foot end (sink) was held at 10 units. Each separate region was then represented by its own morphogen concentration level.

From the experiments, it was possible to work out the period of time needed for regions to change type. This was done by removing the head, and after varying times removing the *1* region and grafting a head on to it. If two sets of tentacles were subsequently formed from the reconstituted *Hydra*, then the *1* region had been determined to form a head. At 18 °C, the 50 per cent success rate for two-head formation occurred at 13 hours. This figure was used in the model to provide data from which the diffusion constant in the model would be calculated. The model '*Hydra*' was then tested to see how it would behave when 'subjected to' the grafting experiments described above. The results obtained from the simulation indicated that the diffusion theory indeed fitted the observed model data. The model is clearly quite simple, and although it suggests that diffusion may be a possibility in producing the regulative pattern, Wolpert admits that it 'should be taken as no more than making diffusion a plausible mechanism' (Wolpert *et al.*, 1972).

The experimental approach in both practical and computer studies has been previously emphasized, when gradients in the insect egg were described (p. 76). The German workers Gierer and Meinhardt (1972) have attempted to put flesh on the bones of the basic gradient model by proposing a *mechanism* for the gradients in terms of biochemical kinetics. The mathematical basis of their model was discussed in Chapter 5. In addition, computer simulations of the gradient mechanism under perturbed conditions have shown the model to be remarkably accurate. In order to simulate the insect gradient, the model proposes that the interaction of activator and inhibitor (p. 58) leads to high activator concentration at the posterior pole of the egg. The inhibitor is then produced largely in this activated region, diffuses into the rest of the egg, and acts as a 'morphogen' forming a concentration gradient supplying positional information to the cells.

Figure 8.15 shows how well the model simulates experimental observation made on insect embryos (Meinhardt, 1977). Meinhardt's work has a significant advantage over other gradient models in that it does provide a biochemical mechanism for the gradient, and so 'an understanding of the interactions of the two substances could help significantly in the design of appropriate assays'. This basic model system has

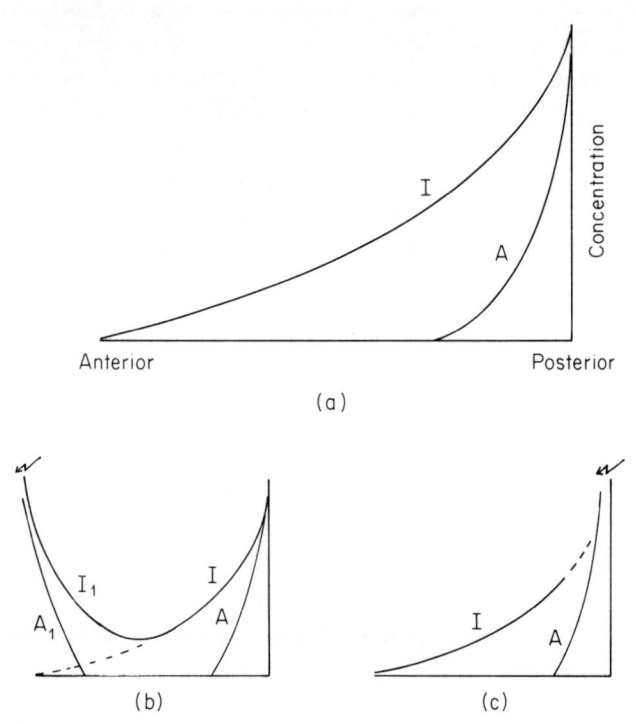

Figure 8.15 (a) Gierer and Meinhardt's gradient model to explain how patterns are specified in the insect egg. Activator (A) is produced autocatalytically at the posterior pole, stimulating inhibitor (I) which diffuses away, preventing other activation centres forming. (b) The model shows why irradiation at the anterior end produces mirror-image structures–irradiation destroys inhibitor and allows a local activator peak A_1 to form. This stimulates a new inhibitor source setting up a double gradient. (c) Posterior irradiation lowers the inhibitor concentration producing a temporary overshoot in activator. No new source is set up, and the prediction is that development should proceed normally. This is the case in the real system.

been applied to a variety of other pattern formation processes: examples are the regeneration of *Hydra* heads (Gierer and Meinhardt, 1972) and formation of veins in plants.

Vertebrate pattern formation has been simulated by Wilby and Ede (1975), who suggested that the pattern of skeletal cartilage differentiation in the vertebrate limb may be set up by a gradient mechanism, using only localized cell–cell interactions. Their theory was that cells modified their metabolism irreversibly at critical threshold levels of a diffusible morphogen which could be synthesized or destroyed by every cell in the system. The simulation may be described by a series of rules

1. Cells are sensitive to the internal concentration of a freely diffusible morphogen, M.
2. At concentrations of M below a lower threshold T_1, cells are inactive.

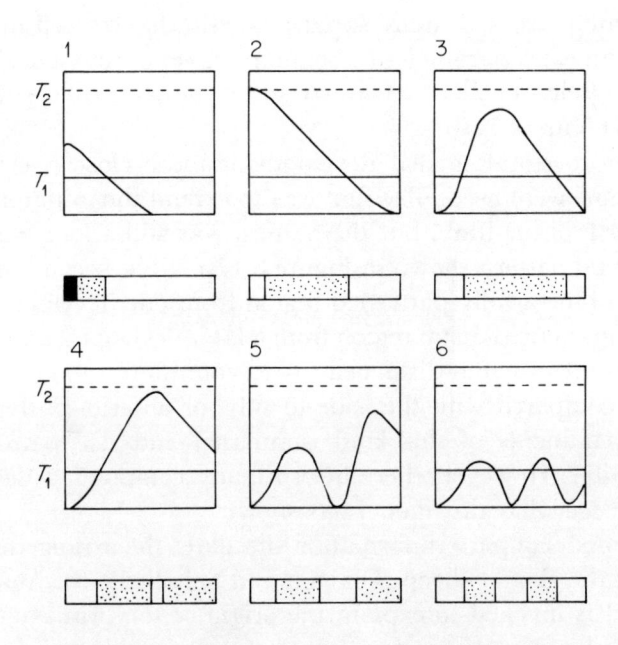

Figure 8.16 Generation of a one-dimensional pattern. Synthesis of the morphogen begins (1) at the left-hand side (shown solid). When the concentration exceeds T_2, cells destroy M (shown shaded). Cells between T_1 and T_2 produce morphogen, and cells below T_1 are inactive. A stable pattern forms (6) because morphogen destruction is irreversible.

3. At concentrations of M above T_1, cells synthesize M.
4. At concentrations of M above a higher threshold T_2, cells actively destroy M.
5. The transformations (2) → (3) and (3) → (4) are irreversible.

By using a suitable diffusion equation, the gradient may be simulated in one or two dimensions. Figure 8.16 shows the type of one-dimensional pattern that may be generated using these rules. An 'initiator region' at the left-hand side starts synthesizing morphogen, which diffuses along the axis where other cells respond to it. A stable gradient eventually forms.

In two dimensions, a more complex pattern forms. To simulate growth on the computer an eight-neighbour space (see p. 85) was used, and Figure 8.17(a) shows the type of simple pattern formed. The cells actively destroying M (i.e., threshold above T_2) are considered to be 'differentiated' areas. If more sophisticatedly shaped 'cell masses' were used, and an extra parameter (a boundary layer of cells actively destroying M − ectoderm cells) was added, more complex 'differentiation' was achieved (Figure 8.17(b)).

In order to simulate real development more closely, several other modifications were used. The first was to extend the initiator region the whole length of the limb, but the pattern was still a long way from the real limb bud pattern shown in Figure 8.17(c). The second modification was to combine a fully patterned region from one developmental stage with an unpatterned distal region from a later developmental stage. This means that the proximo-distal cell-to-cell communication in the model is very low, compared with the 'side to side' or anterio−posterior values. Using amendments of this kind, simulated and real patterns looked rather similar. However, there are so many computer 'rules' that it is difficult to see what this model has shown.

A final model of pattern formation simulates the arrangement of wool follicles in the skin of sheep (Claxton and Scholl, 1973). Again, a diffusion model is invoked to explain the origin of the primary follicle patterns, and the simulation shows patterns similar to those seen *in vivo*. Models have also been produced to investigate the expected sizes of patches of cells in mammalian chimaeras, given an initial ratio of number of cells of each type (Ransom, 1974, described in West, 1974), but these models are computationally very simple and will not be discussed here.

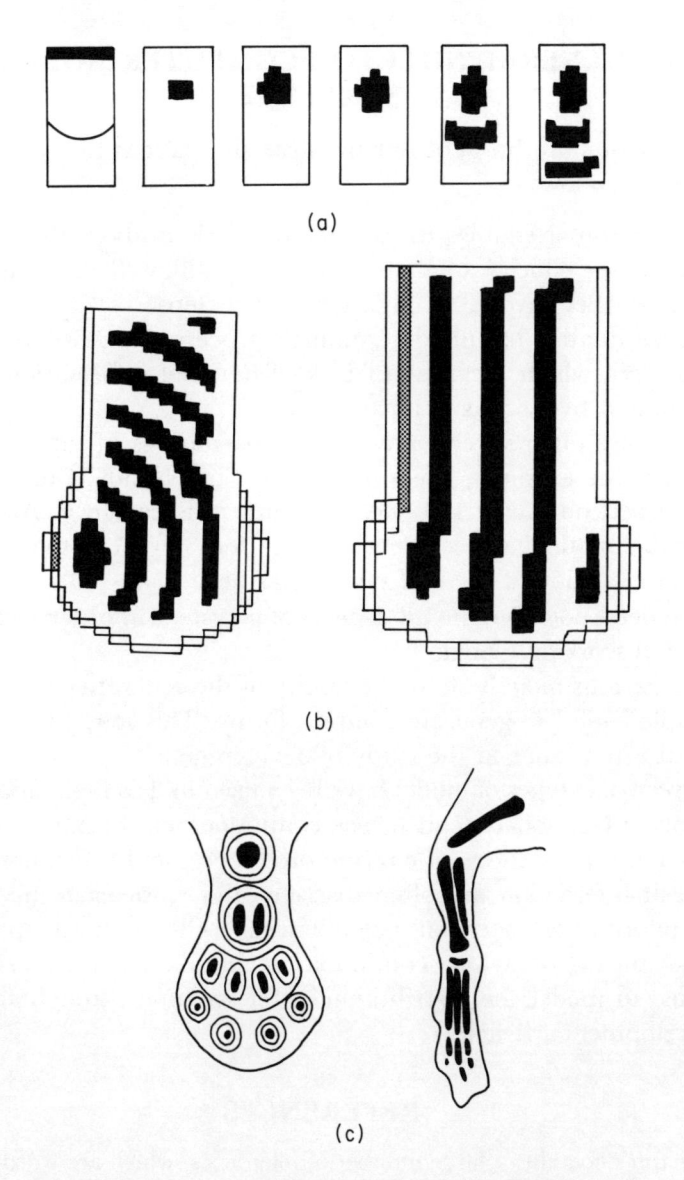

Figure 8.17 A model generating the pattern of cartilage elements in the embryonic chick limb. (a) Development of the model presented in the previous figure in two dimensions. (b) Modification of the model using (left) small initiator region and (right) whole-edge initiator region. The initiator region is stippled. (c) Pattern of cartilage in chick leg early and late in development.

SOME GENERAL NOTES ON COMPUTER MODELLING STRATEGY

Several points are brought out by these simulations and may be summarized as follows.

1. The computer enables us to make predictions about the outcome of hypotheses which cannot be adequately followed through mentally because they involve too many complex steps.
2. The simulation of fully deterministic processes is possible, as well as processes which, because of lack of detailed information, are best simulated by stochastic means.
3. There are often several physical manifestations of the same formal model; for example, the similarity of Ede's and Mitolo's models, described on pages 128–132, falls into this category. Also, once a suitable model has been developed, it does not mean that the *physical* mode of action of the real system has been found.
4. One need not simulate all aspects of a system in order to produce a helpful working model.
5. The various models show the ability of the reiterative application of simple 'rules' to generate complex forms. This may be a method of great importance in the study of development.
6. Three basic types of model have been used in developmental biology. Subcellular models tend to use continuous mathematics, especially differential equations of enzyme production and utilization. Models of cell interaction are often discrete, and emphasize the exchange of information between neighbouring cells. Pattern specification mechanisms often use continuous techniques, like diffusion equations, to model the distribution of information throughout discrete developmental fields.

REFERENCES

This chapter contains a large number of references, which are worthy of study in their own right. They have been separated into the same sections as found in the main text.

Reviews of developmental models

Apter, M. J. (1966) *Cybernetics and Development*, Pergamon Press, London.
Rosen, R. (1968) Recent developments in the theory of control and regulation of cellular processes, *International Review of Cytology*, **23**, 25–88.
Rosen, R. (1972) Morphogenesis, in: *Foundations of Mathematical Biology*, **2**, 1–77, Academic Press, N.Y.

Weiner, N. (1948) *Cybernetics, or Control and Communication in the Animal and the Machine*, Wiley, N.Y.

Subcellular models

Wright's book is reasonably laid out, as is Heinmets's book. The most interesting paper to anyone interested in applying the ideas of self-reproducing automata to subcellular processes is that of Stahl.

Goodwin, B. C. (1970) Biological stability, in: *Towards a Theoretical Biology*, vol. 3, *Drafts* (C. H. Waddington, Ed.), Edinburgh University Press.
Heinmets, F. (1970) *Quantitative Cellular Biology*, Dekker, New York.
Lindenmayer, A. (1971) Cellular automata, formal languages and developmental systems. IVth International Congress for Logic, Methodology and Philosophy of Science, Bucharest, Rumania.
Stahl, W. R. (1967) A computer model of cellular self-reproduction, *Journal of Theoretical Biology*, **14**, 187–205.
Wright, B. E. (1973) *Critical Variables in Differentiation*, Prentice-Hall, New Jersey.
Ycas, M., Sugita, H., and Bensam, A. (1965) A model of cell size regulation, *Journal of Theoretical Biology*, **9**, 444–70.

Automata-theoretic approaches

Von Neumann's book is a classic and is vital background reading for the reader who wishes to sample the flavour of thought at its most original. Arbib's book is less ambitious, but applies von Neumann's ideas directly to developing systems. The papers by Ulam, Eden, and the paper on Conway's work are very readable (especially the latter, which appeared in the 'Mathematical Ideas' section of *Scientific American*). Gordon's paper is a bit of an enigma. How did this way-out paper come to appear in such a conservative and experimentally oriented publication? Perhaps we will never know!

Apter, M. J., and Wolpert, L. (1965) Cybernetics and development, *Journal of Theoretical Biology*, **8**, 244–257.
Arbib, M. A. (1969) *Theories of Abstract Automata*, Prentice-Hall, Englewood Cliffs.
Baker, R., and Herman, G. T. (1972) Simulation of organisms using a developmental model. 2: The heterocyst formation problem in blue green algae, *International Journal of Biomedical Computing*, **3**, 251–67.
Braverman, M. H., and Schrandt, R. G. (1967) Colony development of a polymorphic hydroid as a problem in pattern formation, *General Systems*, **12**, 39–51.
Codd, E. F. (1968) *Cellular Automata*, Academic Press, N.Y.
van Dalen, D. (1971) A note on some systems of Lindenmayer, *Mathematical Systems Theory*, **5**, 128–140.
Eden, M. (1960) A two-dimensional growth process, in: *Proceedings of the Fourth Berkeley Symposium on Mathematical Statistics and Probability*, **4**, 223. University of California Press.

Gardner, M. (1970) The fantastic combinations of John Conway's new solitaire game 'Life', *Scientific American*, **223**, 120–123.
Gordon, R. (1966) On stochastic growth and form, *Proceedings of the National Academy of Sciences (USA)*, **56**, 1497–1504.
Lindenmayer, A. (1967) Mathematical models for cellular interactions in development, Parts I and II, *Journal of Theoretical Biology*, **30**, 455–484.
Maruyama, M. (1963) The second cybernetics: Deviation-amplifying mutual causal processes, *American Scientist*, **51**, 164–179.
von Neumann, J. (1966) *The Theory of Self-Reproducing Automata* (A. W. Burks, Ed.), University of Illinois Press, Urbana, Illinois.
Raven, C. P. (1964) Mechanism of determination in the development of gastropods, *Advances in Morphogenesis*, **3**, 1–32.
Ulam, S. (1962) On some mathematical problems connected with patterns of growth figures, *Proceedings of the Symposia in Applied Mathematics*, **14**, 215–224.

Cell interaction models applied to specific systems

The following models are basic background reading for any serious student of the use of computer models in developmental biology. Some of the most useful papers are those by Honda, Ede and Law, and Williams and Bjerknes. The papers by Goel's group on cellular re-aggregation are also important.

Antonelli, P. L., Rogers, T. D., and Willard, M. A. (1973) Geometry and the exchange principle in cell aggregation kinetics, *Journal of Theoretical Biology*, **41**, 1–21.
Braverman, M. H., and Schrandt, R. G. (1967) Colony development of a polymorphic hydroid as a problem in pattern formation, *General Systems*, **12**, 39–51.
Düchting, W. (1978) A model of disturbed self-reproducing cell systems, in: *Biomathematics and Cell Kinetics* (A.-J. Valleron and P. D. M. Macdonald, Eds.), Elsevier, Amsterdam.
Ede, D. A., and Law, J. T. (1969) Computer simulation of vertebrate limb morphogenesis, *Nature*, **221**, 244–248.
Goel, N. S. *et al.* (1970) Self sorting of isotropic cells, *Journal of Theoretical Biology*, **28**, 423–468.
Gordon, R., Goel, N. S., Steinberg, M. S., and Wiseman, L. L. (1972) A rheological mechanism sufficient to explain the kinetics of cell sorting, *Journal of Theoretical Biology*, **37**, 43–73.
Honda, H. (1973) Pattern formation of the coenobial alga *Pediastrum biwae* Negoro, *Journal of Theoretical Biology*, **42**, 461–481.
Leith, A. G., and Goel, N. S. (1971) Simulation of movement of cells during self-sorting, *Journal of Theoretical Biology*, **33**, 171–188.
Matela, R. J., and Fletterick, R. J. (1979) A topological exchange model for cell self-sorting, *Journal of Theoretical Biology*, **76**, 403–414.
Matela, R. J., and Fletterick, R. J. (1980) Computer simulation of cellular self-sorting: a topological exchange model, *Journal of Theoretical Biology*, **84**, 673–690.
Mitolo, V. (1971a) Un programma in FORTRAN per la simulazione

dell'accrescimento e della morfogenesi, *Dall Bollettino della Societa' Italiana di Biologia Sperimentale,* **47**, 310–312.

Mitolo, V. (1971b) L'abozzo dell'ala nell'embrione di pollo: accrescimento e forma nei primi stadi di svilluppo, *ibid*, **47**, 634–637.

Mitolo, V. (1973) A model approach to some problems of limb morphogenesis, *Acta Embryologiae Experimentalis,* 323–230.

Raven, C. P. (1968) A model of pre-programmed differentiation of the larval head region in *Limneae stagnalis, Acta Biotheoretica,* **18**, 316–329.

Raven, C. P., and Bezem, J. J. (1971) Computer simulation of embryonic development, *Proceedings of the Koninklijke Nederlandse Akademie van Wetenschappen,* Series C, **74**, 209–33.

Valleron, A.-J., and Frindel, E. (1973) Computer simulation of growing cell populations, *Cell and Tissue Kinetics,* **6**, 69–79.

Verdonk, N. H. (1968) Morphogenesis of the head region in *Limnea stagnalis,* Doctoral Thesis, University of Utrecht, The Netherlands.

Vasiliev, A. V. *et al.* (1975) in: *Mathematical Models for Cell Rearrangement,* (G. D. Mostow, Ed.), Yale University Press, New Haven.

Williams, T., and Bjerknes, R. (1972) Stochastic model for abnormal clone spread through epithelial basal layer, *Nature,* **236**, 19–21.

The papers listed below are of less general interest, but might be worth consulting in relation to particular simulations in the text.

Bjerknes, R., and Iverson, O. H. (1968) in: *Proceedings of the First International Symposium on Biocybernetics* (H. Drischel, Ed.), Karl Marx Universität, Berlin.

Braverman, M. H. (1966) Studies on hydroid differentiation, II Colony growth and the initiation of sexuality, *Journal of Embryology and Experimental Morphology,* **11**, 239–53.

Curtis, A. S. G. (1962) Pattern and mechanism in the reaggregation of sponges, *Nature,* **196**, 245–248.

Davis, J. S. (1964) Colony form in *Pediastrum, The Botanical Gazette,* **37**, 75–89.

Elton, R. A., and Tickle, C. A. (1971) The analysis of spatial distributions in mixed cell populations: a statistical method for sorting out, *Journal of Embryology and Experimental Morphology,* **26**, 135–156.

Hawkins, A. F., and Leedale, G. F. (1971) Zoospore structure and colony formation in *Pediastrum* Spp., *Annals of Botany,* **35**, 201–211.

Leblond, C. P. *et al.* (1959) Tritiated thymidine as a tool for the investigation of the renewal of cell populations, *Laboratory Investigation,* **8**, 296–308.

Steinberg, M. S. (1963) Reconstruction of tissues by dissociated cells, *Science,* **141**, 401–408.

Townes, P. L., and Holtfreter, J. (1955) Directed movement and selective adhesion of embryonic amphibian cells, *Journal of Experimental Zoology,* **128**, 53–120.

Models of pattern specification

The computer models in this category are rather more uniform than are the cell interaction models, as less diversity of programming ingenuity is needed. The simplest and clearest paper is that by Lawrence (1971), but this does not go into programming details. The papers by Gierer and Meinhardt have the most to offer in terms of sophistication and biological relevance: especially Meinhardt's 1977 paper. A comparison of the methods used to simulate shell patterns by Waddington and Cowe, and by Herman and Liu, is also interesting.

Claxton, J. H., and Scholl, C. A. (1973) A model of pattern formation in the primary skin follicle population of sheep, *Journal of Theoretical Biology*, **40**, 353–367.

Gierer, A., and Meinhardt, H. (1972) A theory of biological pattern formation, *Kybernetic*, **12**, 30–39.

Herman, G. T., and Liu, W. H. (1973) The daughter of Celia, the French flag and the firing squad, *Simulation*, August 1973.

Lawrence, P. A. (1970) Polarity and patterns in the postembryonic development of insects, *Adv. Insect Phys.*, **7**, 197–260.

Lawrence, P. A. (1971) The organization of the insect segment, in: Control mechanisms of growth and differentiation, *Symposia of the Society of Experimental Biology*, **25**, 379–390.

Meinhardt, H. (1977) A model of pattern formation in insect embryogenesis, *Journal of Cell Science*, **23**, 117–139.

Turing, A. (1952) A theory of morphogenesis, *Philosophical Transactions of the Royal Society of London*, B, **237**, 37–72.

Waddington, C. H., and Cowe, R. J. (1969) Computer simulation of a molluscan pigmentation pattern, *Journal of Theoretical Biology*, **25**, 219–225.

Wilby, O. K., and Ede, D. A. (1975) A model generating the pattern of cartilage skeletal elements in the embryonic chick limb, *Journal of Theoretical Biology*, **52**, 199–217.

Wolpert, L., Clarke, M., and Hornbruch, A. (1972) Positional signalling along *Hydra*, *Nature New Biology*, **239**, 101–105.

The other references quoted in this section are as follows:

Crank, J., and Nicolson, P. (1947) A practical method for numerical evaluation of solutions of partial differential equations of the heat conduction type, *Proceedings of the Cambridge Philosophical Society*, **43**, 50–67.

Postlethwait, J. H., and Schneiderman, H. A. (1973) Pattern formation in imaginal discs of *Drosophila melanogaster* after irradiation of embryos and young larvae, *Developmental Biology*, **32**, 345–360.

West, J. (1974) A theoretical approach to the relation between patch size and clone size in chimaeric tissue, *Journal of Theoretical Biology*, **50**, 153–160.

Chapter 9

A Case study: Computer analysis of insect morphogenesis

How does the would-be simulator go about constructing a computer model of a development process from scratch? The essential prerequisite is, of course, a suitable system for simulation, and some guidelines on how to determine whether or not a system is suitable are given in the present chapter. The experimenter needs to know more than the subject of the simulation: it is not enough to wake up one morning thinking 'I'm going to simulate growth of a frog's leg today'. He must also develop a hypothesis explaining how the frog's leg grows, so that the simulation will provide some meaningful information which will allow the testing of his hypothesis.

Let us assume that the experimenter is a little on the naïve side, and suggests as the basis for his model, that each frog's leg grows from one initial 'frog's leg cell' into a whole frog's leg by random cell division in two dimensions, until the number of cells in the adult leg is formed. If this hypothesis is right, then a suitable simulation should yield a facsimile of a frog's leg. The steps in the construction of the 'frog leg' simulation might be as follows.

Stage 1

This consists of the abstraction of the basic elements in the system. In the example given, these would simply be cells and cell division in two dimensions. The more complex the hypothesis, the more elements will be built into the model, and the simulations to be considered in the main part of this chapter will be more complex than this basic example.

Stage 2

This stage comprises the determination of the simulation method to be used. The hypothesis proposed above, is eminently suitable for array-bound simulation, but this may not always be the case: simulation of diffusion gradients may be directly simulated by mathematical equations (see Chapter 8), for example. If this is the case, the general question of which type of computer to use—digital or analog—may also be worth considering.

Stage 3

This stage involves preliminary simulation to get the feel of the system. It will probably be necessary to go back at this stage either to change elements of the simulation or to change the parts of the real system which are being simulated. It might be decided for example, to use video methods of printout (see Appendix 1) rather than the more tedious paper printout.

Stage 4

This comprises the real simulations to give the experimental data. There will probably never be an 'ideal' simulation of a given system, and the experimenter will constantly be in the position of updating the model. The important thing is that a particular set of data is always considered *in relation to the simulation parameters which generated it*. It would be of no use, and would be scientifically fraudulent, to publish details of a model with data obtained from a different version of the same model. In the example considered above, the model would in no form be able to generate a frog's leg facsimile, so nobody would be fooled if the data showed the opposite to be true. With a more complex simulation, the situation might not be so clear cut.

We will now look at a particular simulation of a developmental process which the author has developed over a period of years (Ransom, 1975; 1977). This simulation (or more accurately, *series* of simulations) followed the rough scheme outlined above, and the account will be laid out in the following form. The initial step will be to describe the development of the fruitfly, the organism on which the simulations are based. Next, the compound eye, the system within the organism with which the simulation is more particularly concerned, will be discussed. This leads on to the particular hypothesis which the computer model was designed to test, followed by an account of the model itself. The chapter continues with an investigation of the similarity between data generated by the computer and the real system.

An acid test of a model is to see if it can account for related phenomena, so this aspect will be considered before concluding with a post mortem on the usefulness of the model in general. Much of the background for the model to be considered has been dealt with earlier in the book.

Many of the methods described in Chapter 7 were developed specifically to study morphogenesis in the larval structures of the fruitfly *Drosophila*. This organism has a twofold suitability for analysis of cell interactions during development. Firstly, the adult structures largely form during larval growth as *two-dimensional sheets of cells* called *imaginal discs*. Secondly, it is possible to use genetic techniques to *physically 'mark' cells in development* so that all their progeny may subsequently be observed. By looking at the patterns formed by such marked 'clones' of cells, information about morphogenetic forces occurring during development may be obtained. No apology is made for the amount of biology in this chapter: in order to appreciate the simulations used, it is essential to understand what is being modelled.

THE DEVELOPMENT OF THE FRUITFLY

The fruitfly *Drosophila melanogaster* (Figure 9.1) has a three-stage developmental period divided into embryonic, larval, and pupal sectors.

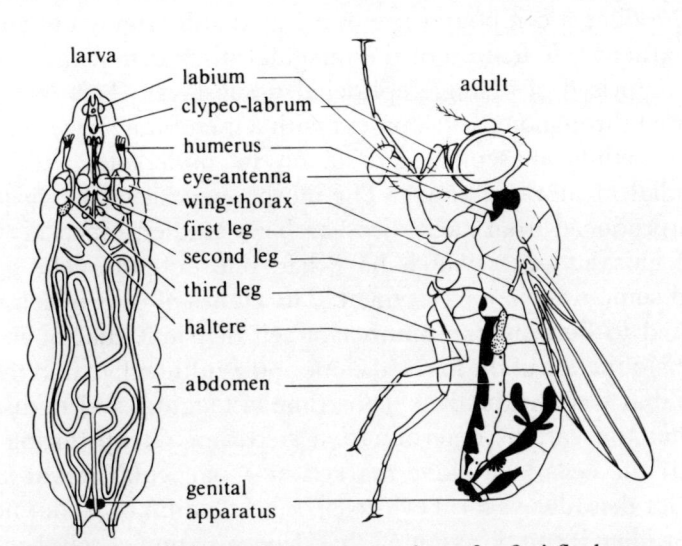

Figure 9.1 Schematic representation of a fruitfly larva (left) and adult fly (right). The positions of the various imaginal discs and the adult structures which form from them are shown.

During the larval stage, cells set aside in the embryo multiply and become determined into imaginal discs, which differentiate during metamorphosis to form the surface structures of the adult insect. There are separate imaginal discs for all the adult ectodermal structures—legs, eyes, genitalia, wings, and so on.

Imaginal discs are especially interesting because, although they seem to become more and more determined throughout larval development, there is no differentiation until the tail end of the larval period (see Chapter 3 for a general description of developmental processes). It is also possible to take out discs from larvae, and to transplant them to the abdomen of host flies or larvae. Disc differentiation occurs when the transplanted disc is allowed to undergo metamorphosis (the change from larva to adult) with the host organism. The availability of the transplantation technique allows us to carry out various experiments to perturb the normal development of the disc, and to see what happens when the disc then undergoes metamorphosis in the host animal. In this way, discs have been broken into small pieces and then transplanted into both younger and older larvae. The results obtained after the subsequent metamorphosis of the treated discs have given valuable information about the developmental organization of disc structures, although, as with other developmental systems, the complex developmental interrelation of the component parts, and the manner in which they cooperate to produce a compound eye or a leg, is still largely unknown.

A characteristic feature of the imaginal discs is that they are all initially comprised of a single epithelial tissue layer. This layer remains associated throughout development with a membrane covering made up of a noncellular basement lamina on its underside, and a cellular 'peripodial membrane' above. The morphogenesis of the various adult organs produced from each disc has been studied especially by using genetic mosaics—flies which have had one or more cells genetically marked some time in development. The clones of cells which result are then used to find out the number of cell divisions that have occurred between induction of the marked clone and adulthood; to find the number of cells that were in the disc at the time of tagging; and to discover the time that the various determination steps are set up during development. If the cells in a clone marked at a particular larval stage in a particular disc always form eye cells, or head chitin cells, but never both together, then it can be assumed that determination of cells has occurred into one of these two developmental pathways.

Besides these quantitative findings, the tracking of a particular cell lineage in this way allows us to analyse the shape of the clone in the adult, and this information can be used to find out the cell division patterns that have occurred during the formation of the clone. Examples of the use of this technique in *Drosophila* are widespread (for example, Postlethwait and Schneiderman (1971) used it to show how morphogenesis occurs in the antenna; Becker (1957) analysed cell division patterns in the eye disc; Bryant (1970) did the same in the wing). These workers have shown both that 'directed division orientations' of cells occur frequently in *Drosophila* larval imaginal discs, and that the boundaries of the marked clones are clearcut—for each larval or embryonic marked cell, there is normally only one patch of cells seen in the adult (Figure 9.2). This observation indicates that there is no clone break-up during development of imaginal discs, and this is probably related to their single-layered nature. Clones in mammalian chimaeras (complex three-dimensional systems) tend to disperse much more easily.

Although the experiments on imaginal discs described above have

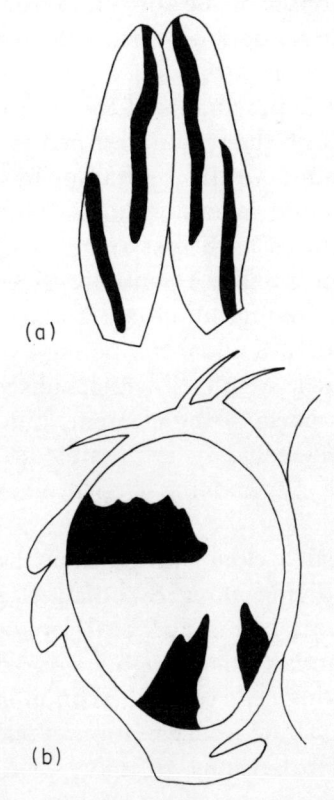

Figure 9.2 Clones in the leg (anterior and posterior aspects) and eye of *Drosophila*.

given us a lot of hard facts about disc development, they do not elucidate any general overall mechanism. For this reason, it was decided that it might be helpful to build a model of the development of some aspects of growth in a particular disc. The disc chosen was that which eventually forms the compound eye and attendant head structures. This disc was chosen because of the wealth of data on clone shapes in the adult eye, and also because the clone shapes suggested that cell divisions in the eye part of the disc take place in certain directions (Becker, 1957).

THE INSECT EYE

In *Drosophila*, two imaginal discs form the compound eyes, head chitin, and head bristles of the adult fly. For our purposes they will be called the head discs. The discs are formed from an invagination of the embryonic pharynx called the 'peripodial sac'. At the beginning of the larval period, about 20 cells are present in each head disc, and this number increases to about 3000 by the onset of the third larval instar. During the third larval instar, eye differentiation starts—the exact time and starter mechanism for this differentiation is not yet clear, but four-cell clusters possibly representing the precursors of the compound eye elements are visible in sections through the region destined to form the eye region 72 hours' development, and eye differentiation has clearly begun by the beginning of pupation and metamorphosis. The presumptive eye cells form goblet-like structures as a first stage of eye formation, and the major part of the process that eventually gives rise to the 30 or so specialized cells of each ommatidium takes place during pupation. Evidence now exists that an individual compound-eye element is not formed from a single determined 'stem cell' which subsequently divides to give rise to the component parts of the element, but from the clustering of unrelated neighbouring cells. Is it possible to find out more about how the eye develops by studying clones marked in the eye during development?

From the shape of such clones in the adult eye, Becker (1957) concluded that the cells divide along two distinct axes during the larval period. The first orientation (Figure 9.3(a)) only occurs up to the end of the first larval instar, and is along roughly a dorso-ventral axis, any 'tilt' depending on the dividing cells vertical position in the eye. At the end of the first instar, a 90° change in orientation takes place (Figure 9.3(b)), and all subsequent cell divisions are more or less along the anterior-posterior axis.

By tracing a large number of the 'twin spot' patterns (twin spots are

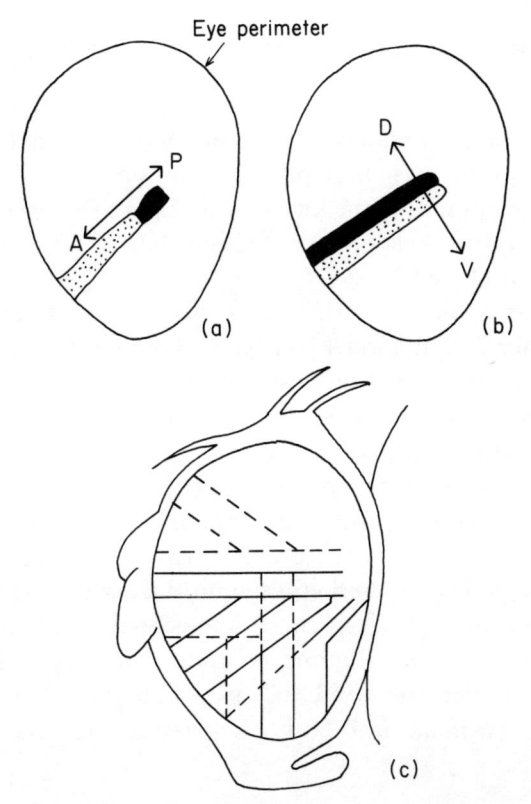

Figure 9.3 (a) Clone produced in the first larval instar. First division(s) occurred along axis A–P. (b) Second instar produced clone, first divisions have taken place along axis D–V. (c) Scheme of division sectors in the eye (redrawn from Becker, 1957).

double clones where all the progeny of both daughter cells of a particular cell division are marked differently, as in Figure 9.3), and superimposing the drawings, it was seen that the clones tended to occur in a number of defined sectors (shown in Figure 9.3(c)), reaching from the posterior to the anterior edge of the eye. As the cells in the twin-spot clones induced prior to the end of the first larval instar (before about 24 hours of larval development) did not divide along the anterior-posterior axis until this time, it was suggested that the concomitant connection made by the developing head discs with the larval brain may somehow result in the new orientation of head disc cell divisions (Becker, 1966).

COMPUTER MODELLING OF INSECT MORPHOGENESIS

The first requirement of a computer model is a list of parameters to be included in the model, and a statement of what the model is supposed to do. The model in this case is required to simulate cell division patterns in the early development of the larval head disc of *Drosophila*; its aims are to throw light on the causal factors behind the cell division orientations seen in the adult clones.

Many of the techniques described in this section are treated in more detail in Chapter 7. The model consisted of a computer program, using as its basic structure a one-dimensional array, with the array elements placed in a two-dimensional hexagonal format. Into the program was embedded a routine calculating the area of an ellipse of given radius. This ellipse was superimposed on to the two-dimensional array, and was 'grown' in steps to represent the membrane surrounding the growing 'head disc'. As the disc itself grows in two-dimensional fashion, and consists of a single-cell layered epithelium during the larval period, the model does not have such great 'dimensional limitations' as those encountered by other array-bound computer models of developmental systems (for example, Ede and Law's two-dimensional computer model of the three-dimensional limb bud, discussed on p. 128).

Running the simulation

At the time each run of the simulation was begun, a variable number of starting cells making up the 'head disc primordium' were placed inside the initial ellipse shape, which took up an area of the array slightly greater than that of the primordium. Each cell was represented on the computer by an array element containing a numeric code differing for each cell type. The ellipse and cell mass were then 'grown' according to the growth parameters chosen; these parameters governed the direction in which the ellipse could 'grow', and the area of the ellipse necessary to be filled with cells before it grew. The ellipse growth increment was also set—incremental growth was necessary because of the discrete nature of the simulation. The one-dimensional array was treated as if it were a two-dimensional hexagonal structure, with 30 columns and 40 rows, as described on page 86.

'Cell division' took place by each cell being 'pointed to' in turn using the list structure to keep track of division order, as described on page 91. All cells had a chance to divide once before a new generation was started, and the picked cell attempted to divide at random into one of the surrounding six sites. If all six locations were filled, the 'pushing' algorithm (p. 94) was used to clear a space. If any of the division directions were constrained (that is, impinged on the growing edge constraint), the cell could not divide in that direction. If all directions were constrained, no division took place, and the next cell to be processed was chosen at random. Cell division was allowed to proceed until about 700 cells were present in the disc, although some runs went up to 2500 cells. Little change in clone position, shape, and relative size occurred between a computer 'disc' size of 730 cells and the roughly threefold increase in size represented by the larger 'disc'. The smaller array was much more economical in computer time, with each run taking one-twelfth of the central processor time of the larger version, and hence many more simulations could be run.

There is no direct evidence regarding the number of eye precursor cells and head capsule precursor cells present at the time 700 cells are in each head disc. Becker (1957) suggests from analysis of clones that the *eye* part of the disc originally derives from 2 cells, whilst Bryant (1970) calculated that there are 13 progenitor cells in the whole head antennal disc complex. A similar technique was used by Postlethwait and Schneiderman (1971), who estimated that there are 7–8 cells in the primordium of the antennal disc. The head disc, if these data are taken together, would start off with 4–6 cells, representing a ratio of between 2/2 and 2/4 cells between eye/head capsule, but such small numbers may well not be very reliable. There are more cells per unit area in the adult eye than in the head capsule, so there must either be more eye precursor cells to start with, or else the division in the eye precursors goes on for longer than in the capsule forming cells. The data does suggest that at the relatively early stage of 700 cells, a minimum of half this number would be determined eye precursors. It is important to emphasize this separation of the head disc into eye and head capsule regions. The computer model was designed to simulate growth of the whole eye/head capsule, and the relative areas assigned to the two types of tissue in the model must therefore be taken into account.

The main problem encountered using the model was that much clone fragmentation occurred (see Figure 9.4). This meant that the large, smooth-edged real clones seen in the published pictures were only observed in a small percentage of the simulations run. The clone break-up was found to be due to separation of cells from their neighbours by the slicing action of other cells dividing, and clearing themselves a free space: if a cell on the edge of a clone happened to be in the line of the slice, it could be separated from the parent clone. Several program

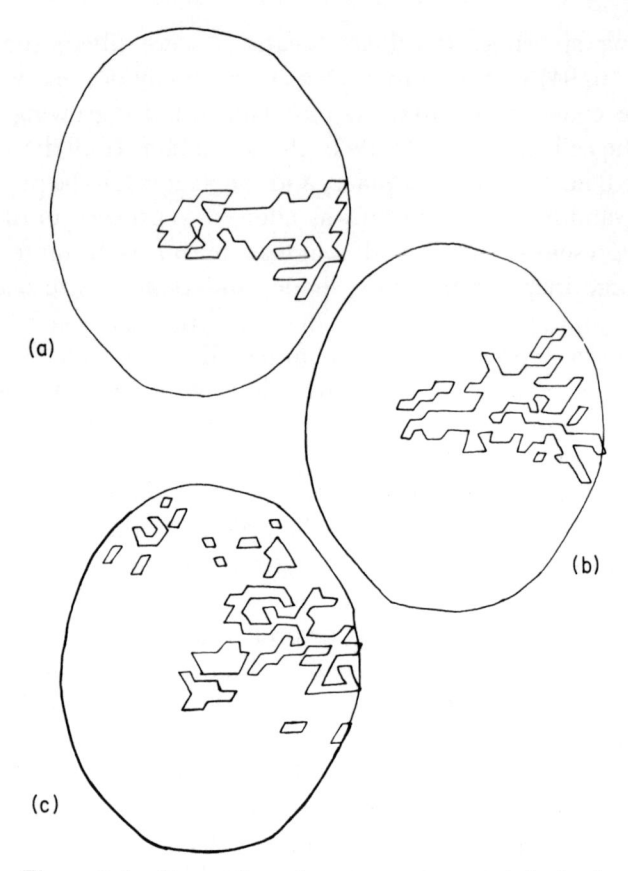

Figure 9.4 Examples of computer-generated single clone patterns produced from the same eight-cell starter primordium. Note the difference between the 'smooth' clone shape of (a) and the 'disperse' character of (c).

variations were employed in an effort to limit this effect, including preferential 'stickiness' between cells in a clone (for example, by making daughter cells not move away from their parents for a certain time), and using complex cells like those described in Chapter 7 to try to buffer the slicing movement between *parts* of cells rather than always needing movement of a whole cell. The latter technique turned out to be very costly in computer time, and 'stickiness' was found to produce inhomogeneities in division patterns: cells further from the edge could often not divide at all because there was so much chance that all division directions would be 'blocked' by sticky cells that could not be moved away satisfactorily.

The 'averaging' technique

Since the fragmented clones were difficult to compare with the clones obtained by experimental means, the following rationale was used. Although the computer clones are dispersed, they still occupy approximately the same region of the cell mass that they would if no cell separa-

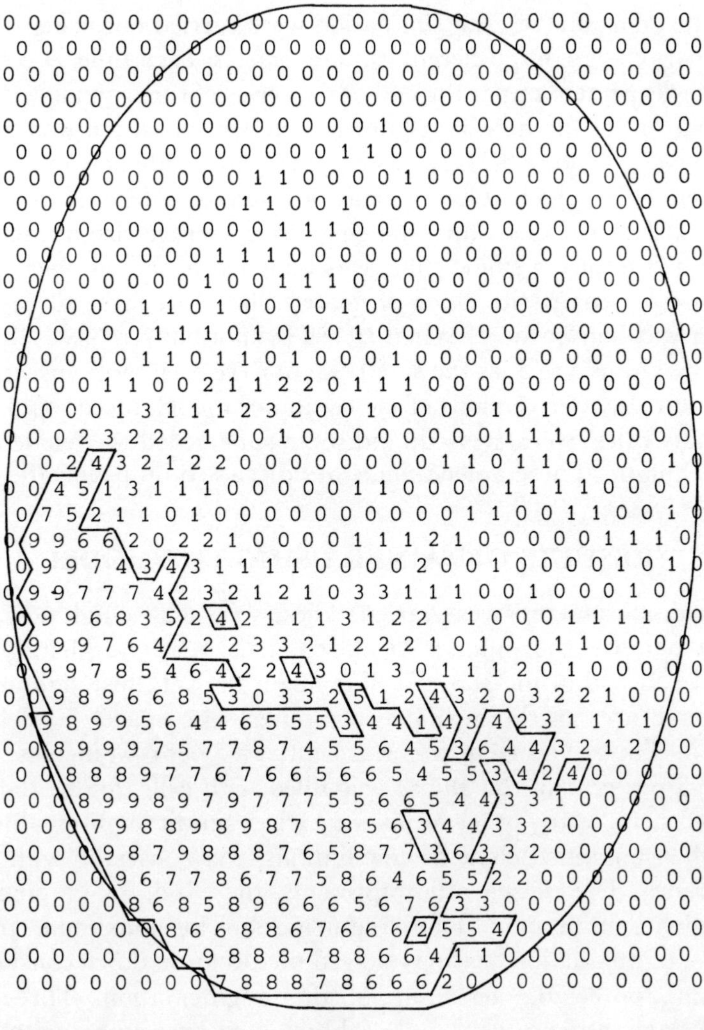

```
        0 0 0 0 0 0 0 0 0 0 0 0 0 0 0 0 0 0 0 0 0 0 0 0 0 0 0
       0 0 0 0 0 0 0 0 0 0 0 0 0 0 0 0 0 0 0 0 0 0 0 0 0 0 0
      0 0 0 0 0 0 0 0 0 0 0 0 0 0 0 0 0 0 0 0 0 0 0 0 0 0 0
     0 0 0 0 0 0 0 0 0 0 0 0 0 0 0 0 0 0 0 0 0 0 0 0 0 0 0
    0 0 0 0 0 0 0 0 0 0 0 0 0 0 0 1 0 0 0 0 0 0 0 0 0 0 0 0
   0 0 0 0 0 0 0 0 0 0 0 0 0 0 1 1 0 0 0 0 0 0 0 0 0 0 0 0
  0 0 0 0 0 0 0 0 0 0 0 1 1 0 0 0 1 0 0 0 0 0 0 0 0 0 0 0
  0 0 0 0 0 0 0 0 0 0 1 1 0 0 1 0 0 0 0 0 0 0 0 0 0 0 0 0
 0 0 0 0 0 0 0 0 0 0 0 1 1 1 0 0 0 0 0 0 0 0 0 0 0 0 0 0
 0 0 0 0 0 0 0 0 1 1 1 0 0 0 0 0 0 0 0 0 0 0 0 0 0 0 0 0
0 0 0 0 0 0 0 0 1 0 0 1 1 1 0 0 0 0 0 0 0 0 0 0 0 0 0 0
0 0 0 0 0 1 1 0 1 0 0 0 0 1 0 0 0 0 0 0 0 0 0 0 0 0 0 0
0 0 0 0 0 1 1 1 0 1 0 1 0 1 0 0 0 0 0 0 0 0 0 0 0 0 0 0
0 0 0 0 1 1 0 1 1 0 1 0 0 0 1 0 0 0 0 0 0 0 0 0 0 0 0 0
0 0 0 0 1 1 0 0 2 1 1 2 2 0 1 1 1 0 0 0 0 0 0 0 0 0 0 0
0 0 0 0 1 3 1 1 1 2 3 2 0 0 1 0 0 0 1 0 1 0 0 0 0 0 0 0
 0 0 0 2 3 2 2 2 1 0 0 1 0 0 0 0 0 0 0 1 1 1 0 0 0 0 0
 0 0 2 4 3 2 1 2 2 0 0 0 0 0 0 1 1 0 1 1 0 0 0 0 1 0
0 0 4 5 1 3 1 1 1 0 0 0 0 1 1 0 0 0 0 1 1 1 1 0 0 0 0
0 7 5 2 1 1 0 1 0 0 0 0 0 0 0 0 1 1 0 0 1 1 0 0 1 0
0 9 9 6 6 2 0 2 2 1 0 0 0 0 1 1 1 2 1 0 0 0 0 0 1 1 1 0
0 9 9 7 4 3 4 3 1 1 1 1 0 0 0 0 2 0 0 1 0 0 0 0 1 0 1 0
0 9 9 7 7 7 4 2 3 2 1 2 1 0 3 3 1 1 1 0 0 1 0 0 0 1 0 0
0 9 9 6 8 3 5 2 4 2 1 1 1 3 1 2 2 1 1 0 0 0 1 1 1 1 0 0
0 9 9 9 7 6 4 2 2 2 3 3 2 1 2 2 2 1 0 1 0 0 1 1 0 0 0 0
0 9 9 7 8 5 4 6 4 2 2 4 3 0 1 3 0 1 1 1 2 0 1 0 0 0 0 0
0 0 9 8 9 6 6 6 8 5 3 0 3 3 2 5 1 2 4 3 2 0 3 2 2 1 0 0 0
0 9 8 9 9 5 6 4 4 6 5 5 5 3 4 4 1 4 3 4 2 3 1 1 1 1 0 0
0 0 8 9 9 9 7 5 7 7 8 7 4 5 5 6 4 5 3 6 4 4 3 2 1 2 0 0
0 0 8 8 8 9 7 7 6 7 6 6 5 6 6 5 4 5 5 3 4 2 4 0 0 0 0 0
0 0 0 8 9 9 8 9 7 9 7 7 7 5 5 6 6 5 4 4 3 3 1 0 0 0 0 0
0 0 0 7 9 8 8 9 8 9 8 7 5 8 5 6 3 4 4 3 3 2 0 0 0 0 0 0
0 0 0 0 9 8 7 9 8 8 8 7 6 5 8 7 7 3 6 3 3 2 0 0 0 0 0 0
0 0 0 0 9 6 7 7 8 6 7 7 5 8 6 4 6 5 5 2 2 0 0 0 0 0 0 0
0 0 0 0 0 8 6 8 5 8 9 6 6 6 5 6 7 6 3 3 0 0 0 0 0 0 0 0
0 0 0 0 0 0 8 6 6 8 8 6 7 6 6 6 2 5 5 4 0 0 0 0 0 0 0 0
0 0 0 0 0 0 0 7 9 8 8 8 7 8 8 6 6 4 1 1 0 0 0 0 0 0 0 0
0 0 0 0 0 0 0 0 7 8 8 8 7 8 6 6 6 2 0 0 0 0 0 0 0 0 0 0
```

TOTAL PICTURE AFTER 10 RUNS

Figure 9.5 Composite ten-run picture. Figures refer to the number of times out of ten that each array position appeared in the marked clone. (For ease of display, '10' appearance has been counted as '9'.)

tion occurred. The *area* occupied, however, is larger than in the latter case, as some edge cells will be widely scattered. If we assume that a clone growing from a cell in the same position in the initial primordium, and subject to the same constraints, is always going to form a clone in the same place in the final cell mass (subject to a certain amount of random variation), then we can use an *averaging* technique to get an idea of the positions which the final clones will occupy. To do this on the computer, ten runs of each simulation were carried out using the same initial parameters, but different random starters each time. A composite picture was printed out at the end of the series of runs, giving a distribution diagram of the number of times each array position was part of a marked clone (Figure 9.5). It was necessary to decide how many 'marks out of ten' were sufficient for an array position to be deemed to be part of the marked clone: to do this, the single run average size of each clone, determined from ten single runs, was matched to the number of markings giving a composite clone of nearest size. In nearly all cases, this number was four or more. When 16 cell primordia were used, the figure had to be lowered to 3, as the small clones were even more susceptible to separation. In later versions of the computer model, more sophisticated 'stickiness rules' were devised, and these superseded the 'clone averaging' technique. These amendments are discussed on page 178.

RESULTS OBTAINED USING THE MODEL

Four main parameters were varied. These were ellipse tightness, ellipse growth pattern, the number of starter cells, and the identity of the marked cell within the starter cell configuration. Ellipse tightness (the percentage area of the available computer 'disc space' filled by cells before the ellipse was enlarged) was kept at a 5 per cent increase in area when 95 per cent of the ellipse was filled with cells, for all the results recorded here. Many other ratios were tried, but little observable effect, other than amplified clone fragmentations, were observed with 'looser' membranes. The chosen ratio represents, therefore, the parameter settings giving minimum clone fragmentation for maximum program speed—'tighter' ellipse settings slowed the program down considerably, with no appreciable lessening of the fragmentation. Three ellipse growth patterns were used: side (abbreviated here to S), with growth totally from left to right'; centre (C) with growth radiating outwards from a point; and an intermediate (I) pattern. These growth patterns are shown in Figure 9.6.

Starting primordia consisting of 2, 4, 8, and 16 cells were simulated.

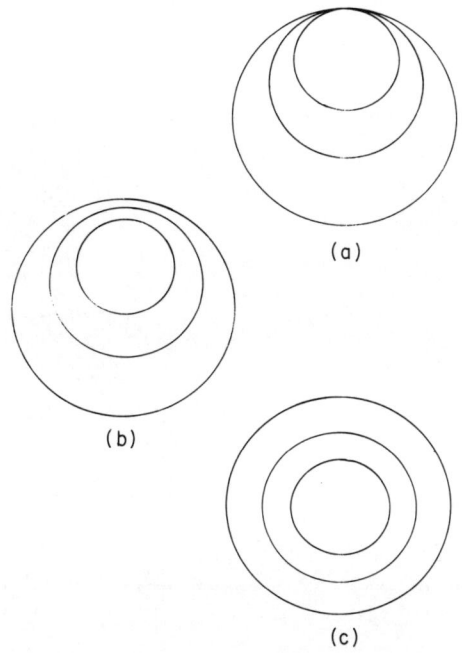

Figure 9.6 Ellipse constraints used to model growth of the insect eye: (a) S growth, (b) I growth, (c) C growth.

All possible marker-cell positions in these primordia were run in the S pattern. Observations on the real system suggest that pushing the growing clones from the posterior would give the closest clone patterns to those seen *in vitro*, and the S results were therefore used as a yardstick with which to compare the results obtained from the C and I growth patterns. The results obtained for each marked starting position consisted of ten individual runs and a ten-run composite picture, made up using the averaging technique.

The computer printout was used in four ways. The growth pattern of the marked clone in each individual run, the final shape and size of each clone, and the composite picture were analysed: the 'twin spots' made by looking at the composite pictures of clones produced by adjacent primordial marker cells were also examined. The growth of each clone could only be observed in those cases where printouts of the clone in various stages of growth were obtained. As this would have resulted in too much output to analyse, not all runs were treated in this way. A version of the computer program was written which ran on a PDP-15

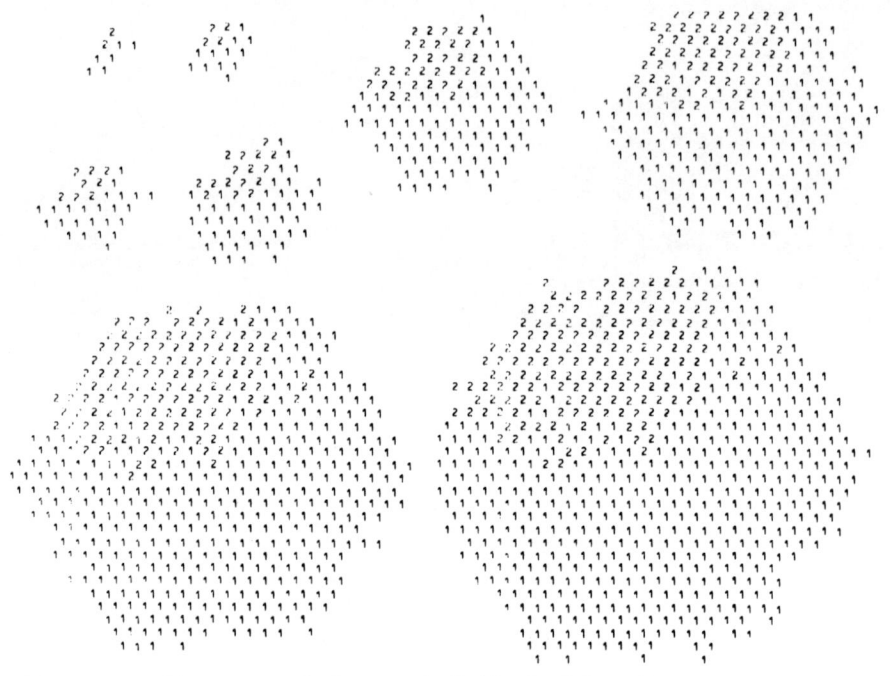

Figure 9.7 Clone growth for an individual eight-cell starter primordium with S growth pattern. 1 = unmarked cell, 2 = marked cell.

computer with video output, enabling the operator to watch the whole developmental process as it occurred. Figure 9.7 shows the progressive growth of a clone derived by marking one cell of an eight-cell starting primordium with no ellipse constraint.

We will now look at the actual output obtained. In all cases, the clone outlines have been drawn onto the computer output, traced, photographically reduced, and then redrawn.

Average clone patterns produced by S growth pattern

Two-cell starters

The final clone filled half of the available area with a roughly straight line separating marked clone from unmarked clone. The orientation of the dividing line varied with the relative orientation of the starter cells. If the two cells were placed with one directly anterior to the other, clones like that of Figure 9.8(a) were produced. If the cells were placed with one above the other, the clone shapes of Figure 9.8(b) were seen.

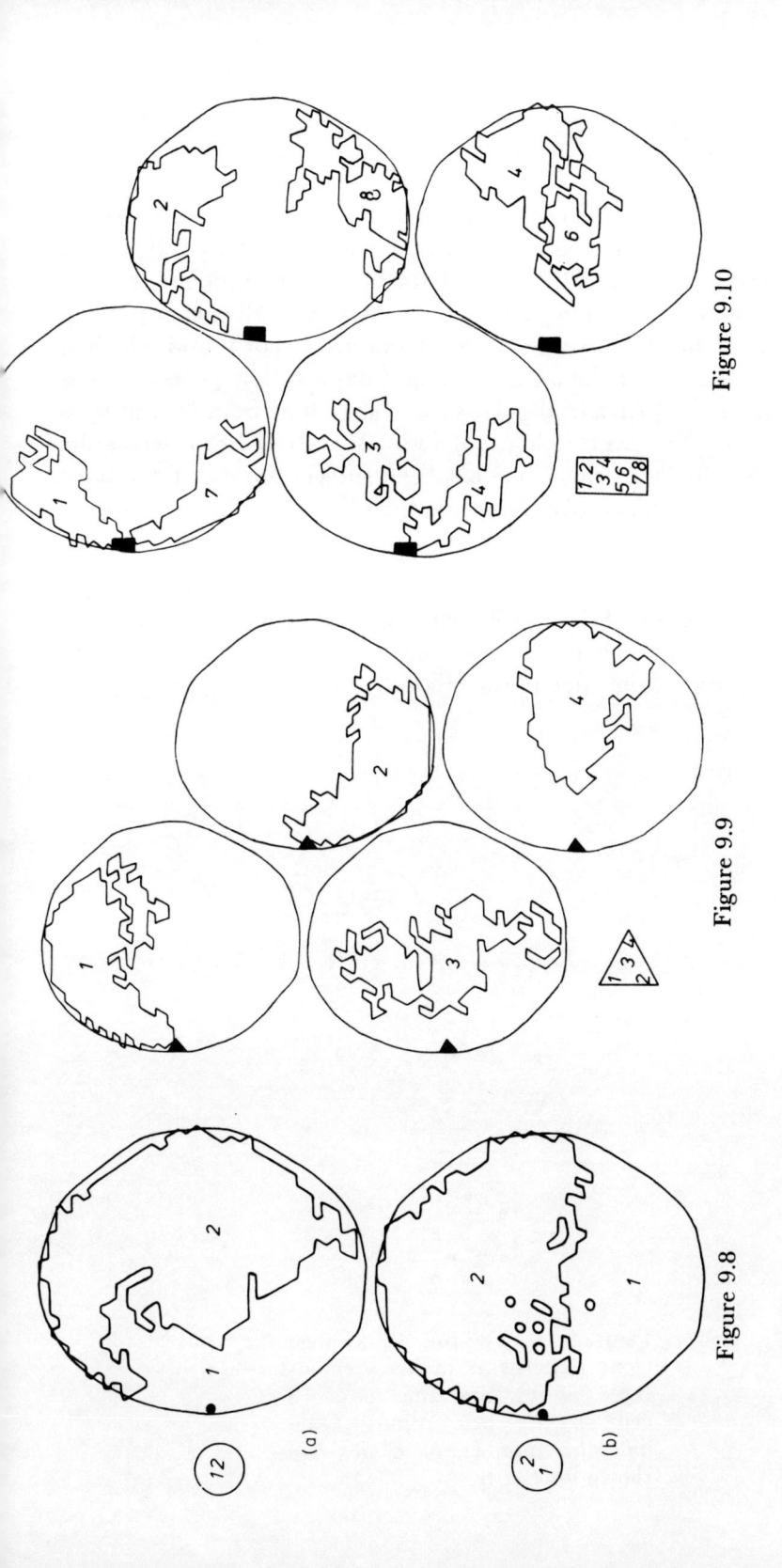

Figures 9.8–9.10 Ten-run average clone patterns with side (S) growth for 2, 4 and 8 cell primordia respectively. The insert figures show the positions in the starting primordia of the cells which eventually formed the numbered marked clones. The starting primordia positions are shown superimposed on the final clone pictures as black shapes.

Four-cell starters

The clone shapes here differed, depending upon the positional relationship of growing clone to the ellipse and to the surrounding cells not in the marked clone. Figure 9.9 shows the starting configuration of the four initial cells, and the positions of the final clones, with all clones growing from left to right. The two posterior clones are swept round with the growing ellipse, whilst the anterior clone (marked 4) is pushed across the growth area to form a triangular section stretching from the centre to the front edge. The centre clone (3) forms a vertical band across the centre of the 'disc'—as this clone has been subject to most disruptions from 'pushings', it has broken up more than the other three clones.

Eight-cell starters

In these clones (Figure 9.10), a difference appears between clones which stretch across the growth area (clones, 3, 4, 5, and 6) and the remainder, which sweep round the edge of the ellipse.

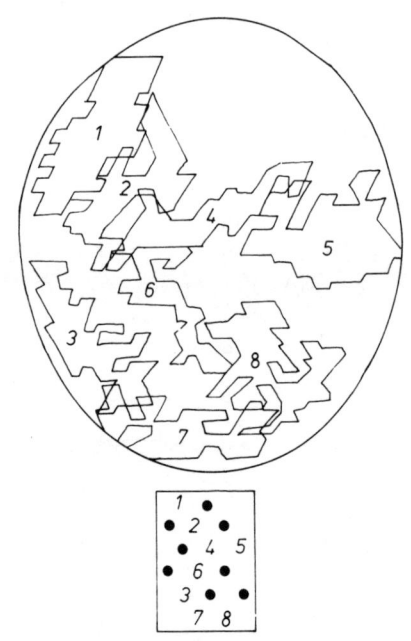

Figure 9.11 Ten-run average clone patterns as in the previous three figures, this time for a 16 cell starting primordium. Only half the number of clones are shown for clarity.

Sixteen-cell starters

The sixteen cell clones shown in Figure 9.11 also show clones which either sweep round the edge of the ellipse, or are pushed across the interior of the ellipse space.

Average clone patterns produced by I and C growth

Eight-cell starters

The tendency for clones to curve round the edge of the growing ellipse was lost by changing the growth direction. Figure 9.12 shows the four

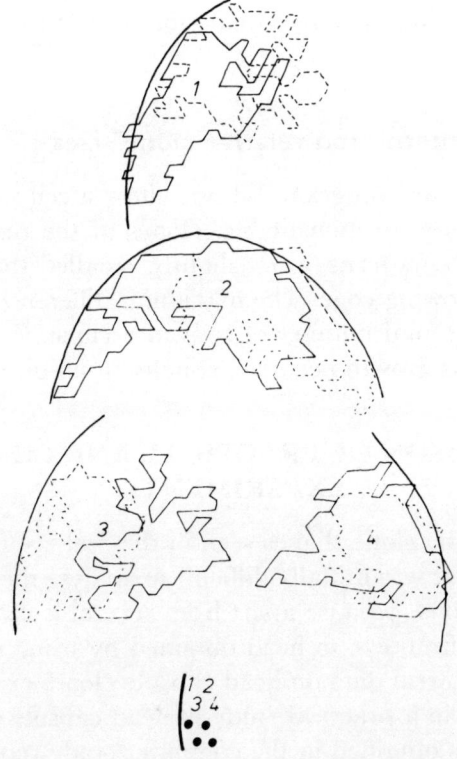

Figure 9.12 Average ten-run clone patterns–side (S) and centre (C) growth compared. Dotted lines show C, solid lines show S growth for the 4 upper cells of an 8-cell primordium. Solid curved lines represent the edge of the ellipse.

upper clones formed from eight primordial cells for both S (solid lines) and C (dotted lines) growth patterns. Clones (a) and (b) especially demonstrate the difference in clone shape produced when the type of growth is changed.

Individual patterns

As described above, the shape of the individual clones was affected to varying extents by the pushing rule, and the coherence of the clones differed markedly depending on the position of the marked cell in the primordium. As was seen in the four-cell starter S pattern, clones formed in the centre of the growth area were more susceptible to this type of separation. Figure 9.4 showed a comparison between individual coherent, non-coherent and 'average' final clones for an eight-cell starter S pattern.

Clone growth patterns and relative clone sizes

Because the computer program did not allow a cell to divide if all six division directions were unavailable, clones at the rear of the growth area in S growth patterns were slightly smaller than those at the anterior, freely growing edge. The maximum difference seen was about 10 per cent of the total number of cells in a clone. C patterns did not show this effect, as growth occurred equally in all directions.

COMPARISON OF PRACTICAL AND COMPUTER EXPERIMENTS

How similar are the clone shapes seen in the real eye clones and in the computer model? It was initially difficult to analyse the results obtained from the simulations, because apart from several isolated exceptions of clones extending from eye to head obtained by using suitable markers (Becker, 1957), no real data on head capsule clones existed. This situation was rectified in a practical study of head capsule clones (Ransom, 1976). The results obtained in the computer study represent the entire (half) head region, which includes head capsule as well as the eye. The computer clone distributions, therefore, represent more than the eye clone distribution diagram (shown in Figure 9.3). It is estimated that a large section of the centre of the computer printout will correspond to the 'eye region' of the disc considered by Becker (1957). This estimate is based on the relative position and proportion of eye and head capsule shown in previously constructed maps of the mature disc (Ouweneel, 1970), and on the clone data considered on page 165.

If the results of the computer model considered in the last section are compared with the drawings of clones in Figure 9.3, it is clear that the posterior to anterior orientation of the clones is duplicated by the S growth pattern. This suggests that there may be an equivalent directional constraint in the real head disc. The pattern of computer generated clones at the edge of the tissue closely resembles the pattern of clones seen on the head capsule (Ransom, 1976). The total pattern of clone orientations in the centre region closely resembles that seen in the fly eye.

It has been suggested that the change in direction of the twin-spot orientations (p. 162) is produced by a stimulus such as the meeting of brain and head disc at the end of the first larval instar (Becker, 1966). The results reported here indicate that the importance of this union might be simply to reinforce the constraining activity of the membrane already present around the disc. There are several different ways in which such a mechanism might work—either the division patterns before brain-to-head disc linking are completely random, or they are already constrained. As there are so few cells present in the disc at the end of the first larval instar (Becker estimates about 20), it will be very difficult to find out which is the case. Alternatively, the large sizes of the clones marked in the first larval instar may, in living system as well as in computer model, be 'too large' to be aligned with one posterior to the other (that is, the posterior clone will always try to push itself alongside the anterior one—see Figure 9.3). This would indicate that the orientation change seen by Becker is a corollary of the properties of induced clones, and does not have any importance *per se* in disc development.

UPDATING OF THE MODEL: SIMULATING LEG DISC GROWTH

If the eye can be modelled, why not the leg, or the wing? Could computer modelling give information on the differences between imaginal discs as they develop? Before embarking on substantial alterations to the model system to simulate the development of other types of imaginal disc, it was first necessary to take into account the evagination movements which take place in discs at metamorphosis. With the eye, evagination is relatively simple, and merely takes the form of local movements of tissue sheets. Each leg, however, undergoes an extension of the order of about 10 times the diameter of the leg imaginal disc which forms it. How would this complicated three-dimensional movement be adequately modelled?

The modelling strategy used depends on the individual tissue. Because the eye and eye imaginal disc from which it grows are so similar in conformation, it is quite possible to consider the clones seen in the adult eye as occupying the same relative sizes and shapes as they do in the mature imaginal disc. With the leg it is more complicated, and a geometrical transformation is necessary to find out what the 'mature discs' clones would have looked like. Evagination from disc to leg takes place as shown in Figure 9.13(a), with the centre of the disc forming the claw, and the disc's outer perimeter forming the parts of the leg joining to the insect body. In a diagrammatic form, this could be simplified as in Figure 9.13(b), where an important principle is shown. The disc is

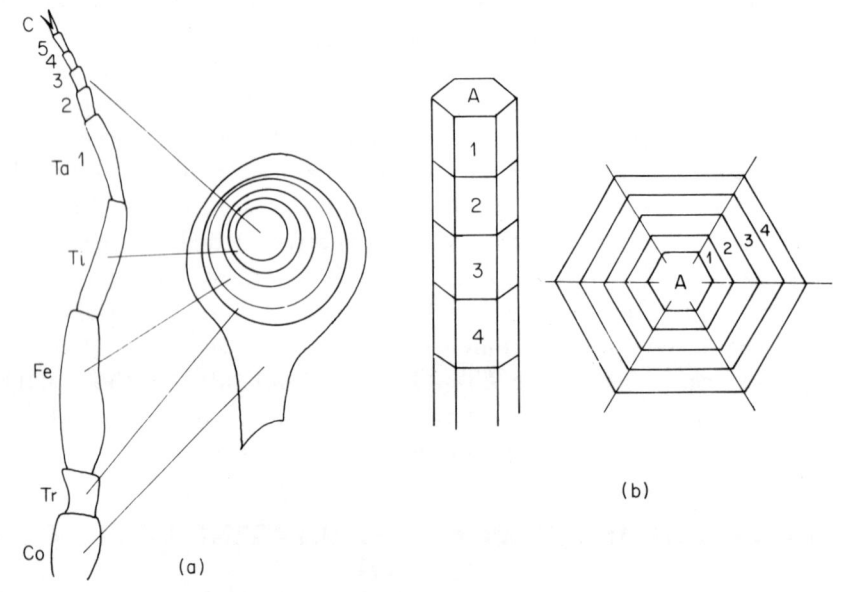

Figure 9.13 (a) Relationship of three-dimensional adult leg and two-dimensional larval disc of *Drosophila*. The positions of homologous parts are indicated. Co = coxa; Tr, trochanter; Fe, femur; Ti, tibia; Ta1–5 tarsal segments; C, claw. (b) 'Metamorphosis' in the computer leg disc. Radial growth produces a roughly hexagonal perimeter shape in the 'disc' and successive perimeters may be drawn through each row of cells as shown (1, 2, 3, 4). 'Evagination' forces may be envisaged as pushing the centre of the disc outwards to form a column representing a stylized 'leg'. Each perimeter is taken to have the same diameter, as an approximation, and clones are mapped onto the 'leg' by estimating the *proportions* of each hexagonal face at each perimeter that is occupied by the clone.

represented as a stack of hexagons. If the stack is drawn upwards out of the plane of the page for a distance much greater than the diameter of the outermost hexagon, to represent the leg, then the rate of change of perimeter diameter is very small over small distances along the 'leg'.

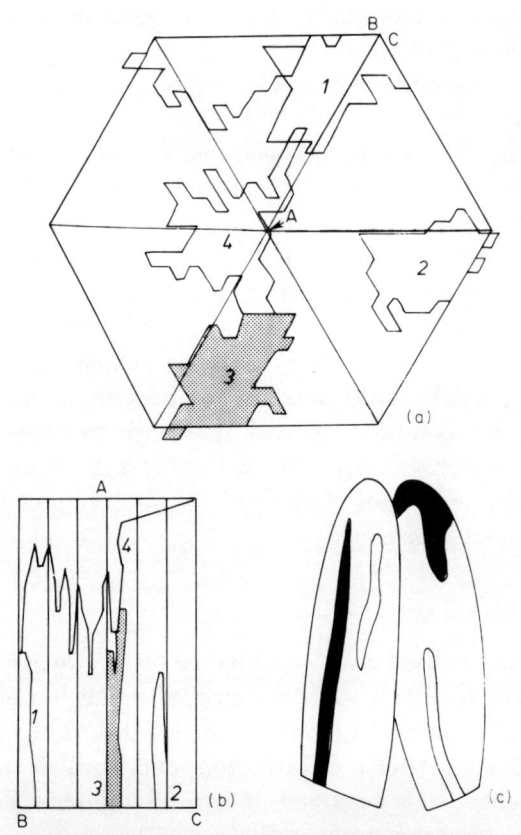

Figure 9.14 Examples of clones using the model described, with the 'nearest free edge' division rule. Computed disc clones (a), their corresponding leg mappings (b), and specimen clones observed on the femur of the adult leg (c) (same orientation as computer leg) are shown. The previous figure shows how computer and real legs may be compared. The computer 'leg' may be opened out as shown here in (b) to exhibit the six faces. After 'evagination', the central point A in the computer disc has the same circumference as the length BC (see legend to Figure 9.13 and text).

This stylistic representation allows us to work out what imaginal disc clones look like from observing adult leg clones. The answer is very simple. Clones on the adult leg are long and thin, running parallel with the length of the leg. The shapes of disc clones necessary to give such adult clones are sectorial, running radially from the centre outwards (Figure 9.14). Similar clones are found all round the adult leg, and this suggests that clone growth in leg discs occurs equally around the disc's circumference. Unlike the eye, there need be no 'constrained' growth in any one direction.

The first change in the original head disc model to model leg growth is therefore simple—the constraining ellipse is removed. Although it could be argued that leg discs also have membranes, their effect is uniform over all parts of the disc circumference. The computer leg simulation should therefore produce clones similar to those obtained using the eye simulation 'C' growth pattern (see above). This amendement successfully produced radial, sector-shaped clones, which were taken to show that the same model would account for growth in differing types of imaginal discs. A second change was also made to improve the model's accuracy, and to answer criticisms of the head disc model: this was to dispense with the 'averaging' technique for observing clone shapes. The method used is described below.

Cells are 'sticky'

The head disc model used an 'averaging' technique to avoid clone break-up, and this was quite rightly criticized as being 'non-biological'. To prevent the fragmentation of clones in the leg disc version, the following rule was introduced. After a cell division, and until the next division of one of the daughter cells occurred, the two daughter cells were 'linked' together in the computer program lists used to represent the cells.

When a cell attempted to divide, the pushing algorithm was used as before, but all pushing movements were checked beforehand to see if separation of any 'linked' cell pairs would occur if that particular pushing movement were permitted. Cells dividing in the center of the growing computer cell mass were more likely to produce separations than were cells dividing on the periphery: an algorithm was therefore devised to control the relative amount of separation allowed, cells in the centre being allowed to cause more separation than the cells on the periphery. A variable N was calculated using the following equation:

$$\frac{\text{distance to nearest free edge (in cell units) } d}{\text{number of separations produced } n} = N$$

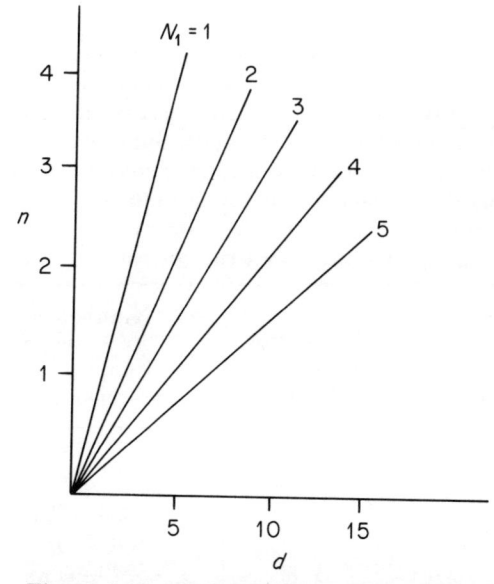

Figure 9.15 Relationship of allowed separations (n) to particular distance to free edge values (d). 'Separation' is defined in the text, and numbers on the lines represent N_1 values. All permitted $d/n = N$ combinations for a given N_1 value (that is, those n and d combinations allowing division to occur) are found to the right-hand side of the line showing the chosen N_1 value.

For any run of the simulation, a value of N, N_1, was set out so that a division was only allowed if N was greater than N_1. This rule allowed central cells to undergo more separations than peripheral cells, and the relationship between the number of separations allowed and the distance to the nearest free edge is shown in Figure 9.15. Table 9.1 shows the relative amount of clone fragmentation, the difference in the number of divisions undergone by central and peripheral cells, and the number of aborted prospective divisions, for N_1 values from 0 to 5.

By using this rule, it was possible to cut out most of the clone separation produced in the earlier head disc model. Figure 9.16 shows how the relative numbers of cell divisions undergone by central and peripheral cells are affected by the value of N_1, and Figure 9.17 shows how clone fragmentation is reduced with increasing values of N_1. It can be seen that the higher the value of N_1, the less divisions are undergone by central

Table 9.1 Clone break-up and other parameters for a specimen clone finally occupying half of the total 'leg disc', as shown in Figure 9.14. The approximate number of cell 'islands' separated from the main clone is shown for N_1 values 0 to 5. The 'maximum division difference' represents the greatest difference in numbers of divisions undergone by cells in the full grown disc (see Figure 9.15). The 'number of aborts' is the number of single division directions, or the number of whole cells, refused division over the course of the whole simulation as a result of the 'stickiness' rule used. In each case, extreme ranges have been recorded, and at least five simulations have been run for each N_1 value.

N_1	Islands	Maximum division difference	Number of aborts Single direction	Whole cell
0	5–10	1	0	0
1	6–10	1	0	0
2	2–12	2	885–1021	83–104
3	4	3	4002–4017	601–608
4	1–2	5	5770–6250	913–992
5	0	7	7675–7687	1239–1241

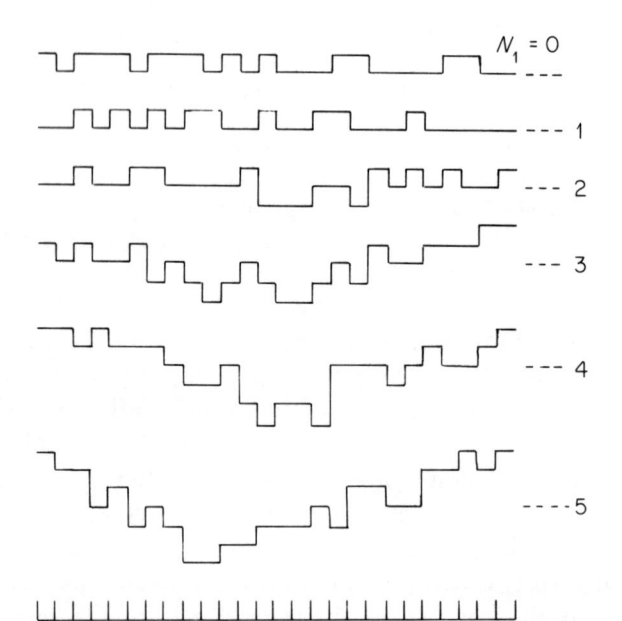

Figure 9.16 Histograms showing cross-sections through a full-grown computer disc of the number of cell divisions undergone by each cell, for values of N_1 from 0 to 5. The dotted lines show nine cell divisions, and cell width is indicated by the graduated scale.

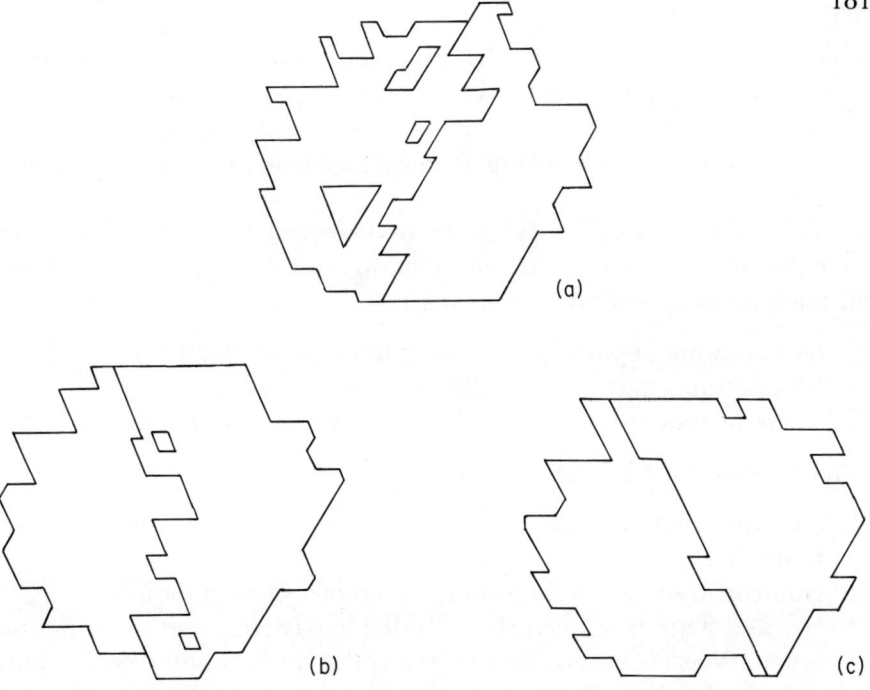

Figure 9.17 Clone boundaries for different N_1 values. A clone occupying half of the total disc space is shown in each case, and the starting positions of the cells in the primordia producing these 'discs' was the same in each case, with two cells being placed alongside one another on the horizontal axis of the page. When N_1 values are high, the cells adhere together more firmly: this has the effect of reducing clone break-ups. (a) $N_1 = 0$, (b) $N_1 = 3$, (c) $N_1 = 5$.

cells. This differential mimics the observed differences in cell division rates seen in *Calliphora* wing and haltere discs (Vijverberg, 1974). A similar difference in cell division rate has been noted in the tye (Becker, 1957) between posterior and anterior regions of the growing eye.

Examples of 'leg clones' obtained using the above modifications with $N_1 = 3$ are shown in Figure 9.14.

POSTSCRIPT

In previously published accounts of the above model (Ransom, 1975, 1977) further discussion of some of the biological implications of the model are presented. How can the model be verified experimentally? What could produce the 'constraint mechanism'? Are constraints of general biological importance during morphogenesis? These are questions which are thrown up by the computer simulation. The most important point to raise here, however, is whether or not it was worth doing the simulation in the first place—do the results suggest new lines of research, fresh ways of looking at morphogenetic processes in imaginal discs?

We could summarize the state of knowledge about cell division activities in discs prior to the information given by the computer model in the form of a series of statements:

1. Divisions may be oriented in imaginal disc development.
2. Mitotic rates may differ in different parts of discs.
3. These factors are probably important in disc morphogenesis.

The computer model suggests that:

1. Oriented divisions may well be produced by constraints ('membranes'?).
2. Non-constrained cell division in two dimensions is radial.
3. If 1 and 2 are true, then the cells divide passively and their division orientations are determined by the whole mass, and not by the individual cells themselves.

It would be very difficult to have put forward a case for investigating this sort of mechanism without first constructing a computer model. This is because before building a model to describe a particular process it is necessary to decide *how* the process can be best simulated. In this case, the only two ways to obtain the necessary clone orientations would have been: (1) to introduce constraints; or (2) to endow each individual cell with a complex apparatus for interpreting some signal (positional information?) telling it how it is orientated. There is no evidence that disc cells do have polarity-setting mechanisms of this kind: a model invoking these properties would probably work, but its involved nature would not suggest it as a strong candidate for an *in vivo* mechanism.

In its own turn, the model gives us sufficient insight into the system for it to be possible to design experiments to test whether, for example, the prediction that clones 'sweep' around the edge of the central eye region is true. This was found to be the case (Ransom, 1976).

REFERENCES

Any reader wanting more background on the development of the *Drosophila* eye should read Becker's excellent review. A general introduction to *Drosophila* development may be found in the *Handbook* edited by the present author.

Becker, H. J. (1966) Genetic and variegation mosaics in the eye of *Drosophila*, *Current Topics in Developmental Biology*, **1**, 155–71.
Ransom, R. J. (Ed.) (1982) *Handbook of* Drosophila *Development*, Elsevier, Amsterdam.

Other sources referred to in the text:

Becker, H. J. (1957) Über Röntgenmosaikflecken und Defektmutionen am Auge von *Drosophila*, *Zeitschrift fur Induktive Abstemmungs und Vererbungslehre*, **88**, 333–373.
Bryant, P. J. (1970) Cell lineage relationships in the imaginal wing disc of *Drosophila melanogaster*, *Developmental Biology*, **22**, 389–411.
Ouweneel, W. J. (1970) Normal and abnormal determination in the imaginal discs of *Drosophila*, with special reference to the eye discs, *Acta Embryologiae Experimentalis*, 95–119.
Postlethwait, J. H., and Schneiderman, H. A. (1971) A clonal analysis of development in *Drosophila melanogaster*: Morphogenesis, determination and growth in the wild-type antenna, *Developmental Biology*, **24**, 477–519.
Ransom, R. (1975) Computer analysis of division patterns in the *Drosophila* head disc, *Journal of Theoretical Biology*, **53**, 445–462.
Ransom, R. (1976) Cell division patterns in the *Drosophila* head disc: clones on the head cuticle. *Journal of Embryology and Experimental Morphology*, **36**, 109–125.
Ransom, R. (1977) Computer analysis of cell division in *Drosophila* imaginal discs: model revision and extension to simulate leg disc growth, *Journal of Theoretical Biology*, **66**, 361–377.
Vijverberg, A. J. (1974) A cytological study of the proliferation patterns in imaginal discs of *Calliphora erythrocephala* Meigen during larval and pupal development, *Netherlands Journal of Zoology*, **24**, 171–217.

Chapter 10

Developmental modelling and the future

We must now look both back at the value of the approaches set out in the preceding chapters, and forward at the ways in which these and related techniques can best be used in the future to analyse developmental processes. The various techniques and methodologies of use in modelling developmental systems are diverse, and no previous attempt has been made to bring examples of most of them into print together. The few existing studies of such models have been under essentially restrictive headings like 'morphogenesis' or 'cellular automata' and have dealt either with the authors' own model system (for example, Arbib, 1969), or with one particular aspect of development (Rosen's (1972) review of theoretical aspects of the 'self-assembly' side of morphogenesis). It is hoped that the chapters of this book provide something of an introduction to the field of models in developmental biology, and so partly rectify the gaps between computer manual, mathematical textbook, and biological model that have existed in the past.

It is necessary, when dealing with a basically theoretical topic, to show that the methods studied can be applied to give useful advances in knowledge, and are not just 'models for the sake of models'. The degree to which this statement is applicable in any individual case depends partly on the type of model to be constructed. A paper-and-pencil model will be within the repertoire of any good biologist, and he will probably make the connection between model and real system at a subconscious level. A mathematical or computer model may require more thought as to the relevance and applicability of the model. This point will be elaborated on in the next section. Substitute system models are less theoretical in nature, but as we have seen in Chapter 4, care is also necessary when extrapolating findings obtained in experiments on such systems to other more complex systems.

Throughout this book the assumption has been constantly made that complex systems require modelling strategies for their solution. The problem is therefore to gaze into a crystal ball to see which types of models will be most useful in the long term. It is tempting to suggest that models of the future will be based around the computer. The digital computer allows simulation of models which allow both qualitative and quantitative components to be built in. Whether or not computer models will be based on heuristic principles or will involve sophisticated mathematical concepts is a separate issue. Mathematicians must certainly expend more effort in making their models accessible to the general biologist in order for their studies to be taken seriously.

FUTURE RESEARCH

So how do we find out what rules govern the development of an organism? Does the material described in this book allow us to make any predictions about the shape that future research should take? With these thoughts in mind, some of the most useful fields of research to tackle in the future can be outlined.

Cells and abstract automata

Further investigation of the similarity between the behaviour of cell systems in biology and the behaviour of abstract automata may yield clues about the generation of ordered patterns during development. An interesting 'clue' that this may be so is provided by the work of Kauffman (1969). Briefly, Kauffman considered a computer simulation of randomly connected two-state 'genes' that may be switched either on or off. If each unit was connected to two others, and a signal was passed randomly and linearly through the connections, it was found that short, stable cycles resulted. In fact, a net of 1000 elements possesses 2^{1000} possible states. The average net cycled amongst only 12 of the states. There is no explanation at present for this result, but the implication here of 'order out of chaos' may be relevant to the stability of living organisms. Kauffman was quick to notice the similarity himself and suggested that 'nets of metabolism' were randomly constructed during the evolution of living organisms. This idea is attractive, because it indicates that primordial organisms may have originally formed from collections of reaction nets built randomly.

Computer techniques

Computer scientists should continue to develop computer language facilities which allow more sophisticated programming to be carried out.

Ideally, it should be possible to produce a language allowing large numbers of cells to be processed in parallel. More efficient storage mechanisms will also be important when, for example, considering cells made up of 'flexible' component parts. Interaction seems an essential aid to computer modelling of developmental processes, and so sophistications in software enabling simpler communication between user and program will be useful. The use of video, especially, is hampered in many installations by having to write (or get written) sections of computer program in machine code to handle, for example, the use of a light pen (see Appendix 1).

Conflict between practical and theoretical studies

Biologists are basically experimentalists, and modelling must be used in conjunction with practical work in many cases. At present there is much mistrust of modelling, particularly that involving the computer, and only a few research groups seem to have achieved much harmony between the two approaches. Model → prediction → experiment → model is the most satisfactory way of tackling many problems in development, and the use of the computer can augment the more limited possibilities of 'pencil-and'paper' models.

The lack of communication between biologists who use computer-aided heuristic models and those who claim that such modelling is a complete waste of time is somewhat difficult to understand. Any biologist has to propose a model in order to provide a working base for his experiments. As we have seen, even a purely descriptive biologist who may only be interested in, say, the structure of the insect compound eye, has to construct a mental model of how the component parts fit together.

Developmental biologists deal with some of the more complex of biological phenomena, and in many cases there is no way that a few simple diagrams will allow an experimenter to understand the system he is studying. For a start, developmental processes involve a time factor, so at the very least, a number of diagrams would be required of the system at different stages of growth. We have already discussed the cellular make-up of most developing systems, and this adds a further degree of complexity. All in all, this adds up to a terrifyingly complicated picture, if consigned to a static piece of paper.

Nevertheless, many biologists refuse to take the logical step and 'animate' their models by using computer or mathematical methods. The mistrust covers three main areas. Firstly, there is the traditional biologist's distaste for anything that smells vaguely of mathematics. This is misfounded, as I hope will have become apparent in the previous chapters. Secondly, there is a mistrust of workers who spend all their time building sophisticated models. This is partly understandable, as such researchers may occasionally be guilty of poaching experimental results, and may even interpret these results subjectively. The problem of subjective interpretation is more rife the less the biological training of the modeller. Thirdly, the success of computer and mathematical modelling has been limited so far: it may well be that this is due to the relatively small number of models constructed and the small number of scientists who are actively engaged in computer modelling of development. The cynics might suggest that lack of success means lack of usefulness, but I hope that the arguments and examples laid out here have at least given food for thought.

Towards a theory of development

It is also difficult to decide on the basic unit of the model. Cells are favoured by many authors (Ransom), but so are fields determined by biochemical signals (Meinhardt and Gierer) or even genetic switches (Kauffman, Slack). By presenting examples of the various types of models here in the same volume, I hope that the choice is made a little clearer to the prospective modeller.

As discussed in the introduction to Chapter 2, the study of biological processes involves modelling many features unencountered in the physical world. The delicate relationship between random and deterministic processes and the hierarchical organization of biological systems are two of the major stumbling blocks to our understanding of the development of living organisms. Will it be possible in the future to describe developmental processes, to present a unifying theory that encompasses everything from regeneration in *Hydra* to the development of neural connections in the human brain? What form would such a theory take?

One certainly is that a verbal model will not do. There are too many features that must be taken into account. The value of the 'theory of evolution' is that it can be summed up, at least for the layman, in the two words 'natural selection'. It is difficult to envisage the popularization of a 'theory of development' in the same terms. By similar criteria, computer modelling is also a dead-end. Computer modelling in developmental biology tends to revolve around the heuristic simulation of individual processes, and although insight into the workings of such processes is given there is no suitable framework on which to construct our developmental theory.

We must therefore turn to mathematics to fulfil this role. The prospective theorist is faced, however, with a bewildering array of mathematical tools. Should he use Thom's catastrophe theory, mathematics based on partial differential equations like Turing and subsequent workers, or an automata-theoretic approach as used by the proponents of Lindenmayer's L-systems.

There are two central problems, the first of which has already been alluded to in this chapter. This is the difficulty of communication between biologist and mathematical modeller. With many branches of mathematics it is just not possible for the biologist to wave a magic wand and instantly comprehend the mathematical arguments used—even in broad terms. Unlike biology, where concepts can be simplified and explained to any competent layman, higher mathematical reasoning requires a firm grounding in the basic concepts—a novice cannot be expected to go from arithmetic to Laplace transformations in a day. The biologist is therefore reliant on being led blindfold by the mathematician, and it is no wonder that many biologists rebel at such a situation.

The second problem is for the mathematicians to solve, and involves the techniques to use in the study of development. The present author has been struck by the shortcomings of both differential equations and catastrophe theory on the one hand, and automata theory on the other, to adequately describe development. The former group of mathematical concepts have no ability to model the discrete aspect of development: the important rule that most developing systems are cellular. Conversely, automata-theoretic models are often almost 'clockwork' in nature, and lack the machinery to deal with continuous biochemical fluctuations in developing tissues. Organisms therefore have both discrete and continuous components, and any general mathematical model must account for both characteristics.

The challenge is therefore laid, and perhaps a new type of 'hybrid mathematics' must be devised to describe developmental processes adequately.

CONCLUDING REMARKS

There are two opposite extremes in developmental biology. On the one hand, many 'right-wing' embryologists cling to the conservative notion that cells differentiate due to the influence of a small number of wonder chemicals. If we knew what these chemicals are, they argue, we would understand how development works. The 'left-wing' argue, however, that development proceeds by complex interactions, and that it is only by model building that the instructions underlying the sociology of cells can be worked out. The conservatives have spent 50 years looking for their wonder chemicals, whilst the modellers mostly started in the 1970s. Only time will tell which approach will be the more fruitful in the long term.

REFERENCES

Arbib, M. A. (1969) *Theories of Abstract Automata*, Prentice-Hall, Englewood Cliffs.

Kacser, H. (1960) Kinetic models of development and heredity, in: *Symposia of the Society for Experimental Biology*, **14**, Models and analogues in biology, 13–27.

Kauffman, S. A. (1969) Metabolic stability and epigenesis in randomly constructed genetic nets, *Journal of Theoretical Biology*, **22**, 437–467.

Rosen, R. (1972) Morphogenesis, in: *Foundations of Mathematical Biology*, **2**, 1–77, Academic Press, N.Y.

Thom, R. (1975) *Structural Stability and Morphogenesis*, Benjamin, N.Y.

Appendix 1

Video techniques

DISPLAY OF OUTPUT

The reader will probably be familiar with the variety of output devices that can be coupled up to a computer. These consist for the main part of line printer, teletype, graph plotter, and video displays. Unfortunately, it may not be easy, especially when writing a new program, to work out the usefulness and time and cost effectiveness of particular devices for particular simulations. The following scheme shows an ordered schedule that was actually used in the running of a simulation program (the imaginal disc model described in Chapter 9). It may help to illustrate how the various output devices may be most profitably used in developing a model of this sort.

1. Model worked out and described (pencil-and-paper).

2. Program written, and preliminary form punched on paper tape.

3. Program fed into multi-access system, and debugged using a teletype with teletype output.

4. Teletype operation for program development, with line printer output for speed.

5. Developed program amended and transferred to video system, using a small dedicated computer. 'Behaviour' of the program directly observed on the video screen.

6. Program run using batch process on large computer for large number of runs.

Video has distinct advantages over hard-copy devices like line printer, graph plotter, or teletype. These latter can only print 'time slice' samples of the progress of a simulation. The video screen, on the other hand, allows us to watch everything that is happening, as changes in array element values can be simultaneously updated on the screen. It is even possible to make cine films of simulated morphogenetic movements.

ADVANTAGES OF VIDEO METHODS

Besides giving greater accessibility of program to programmer, video also facilitates the use of interactive techniques. Using a computer interactively implies that man and machine are in some sort of feedback loop—what the machine produces influences the next set of data fed in by the user. In the most literal sense, almost every use of the computer could be considered interactive, even the act of compiling an incorrect program, which will require re-submission after the user has examined the diagnostic information produced by the compiler. But, to redefine the concept further, we can distinguish the interactive program by its requirement for user intervention during the course of execution, in response to the information it has produced.

It is possible to halt the video program at any point during its running and make changes in the program's execution—for example, cells might be 'grafted' onto or 'removed' from a growing 'cell' population. This will be discussed in more detail below.

There are a large number of video screens available, falling into two main categories: cathode ray tube (CRT) systems (also called refresh displays) working on the same principle as the television screen, and storage tubes. The storage tube has a 'gun' which emits a net positive charge on to the regions written on the phosphor-coated screen—this charge is maintained and seen by the action of a 'flood gun', flooding the screen with a uniform pattern of electrons. With a storage tube, individual points on the screen cannot be changed or overwritten, and the screen has to be wiped afresh before new information can be introduced.

USING A REFRESH DISPLAY WITH A CELL POPULATION GROWTH PROGRAM

This is an easy technique, although just how easy it is depends on the state of the software facilities of the particular computer coupled to the CRT. The main difference between a program used with conventional output, and the program devised for video, is the presence in the video program of a 'display file'. This consists of a modified array, basically similar to the main array of the program, used to mark the positions of the 'cells' (Figure A1.1).

The array is enlarged and modified in order to contain not only cell positions, but also control instructions for driving the video. These instructions have to be inserted into the correct places in the video array, and so they make it an asymmetrical structure. Array positions 1–5 are

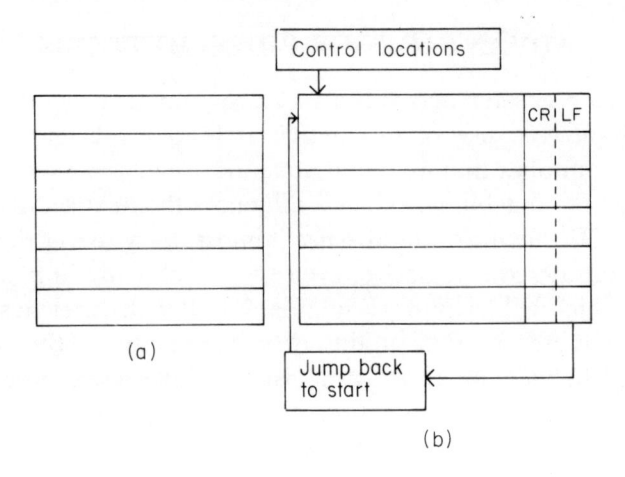

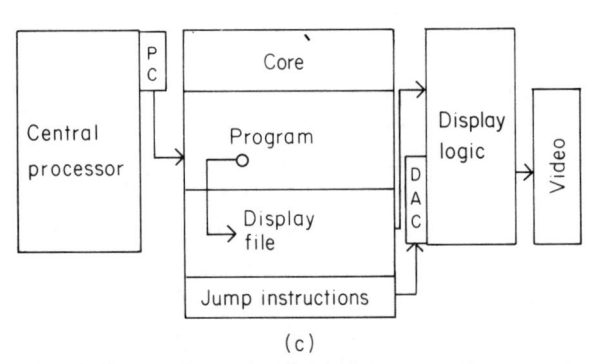

Figure A1.1 The make-up of a video system. (a) Main, and (b) modified video arrays. CR = carriage return, LF = line feed. (c) Computer configuration used for video. The display file picks up information from the modified video array in the program: the display logic transforms this information into the electronic impulses required to drive the video. PC = program counter, DAC = display address counter.

used as control words to keep the refresh display cycling. At the end of each block of elements constituting a 'row' of normal array positions, two extra array positions specify 'carriage return, line feed' for the raster. A final array position at the end of the program makes the display processor (the computer hardware linking display file with screen) loop back to the beginning of the file and repeat the process over again. Meanwhile, array elements in the main array may have changed value,

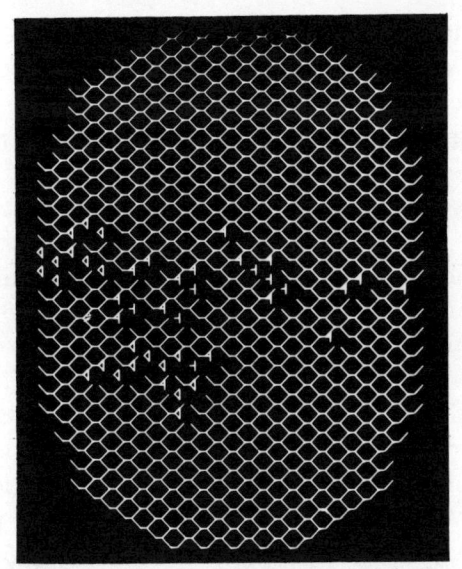

Figure A1.2 An example of an induced clone in a simulated *Drosophila* eye (see Chapter 5). Two special shapes, ⌣ and ◁, have been defined in subroutines, so that a more pleasing picture can be produced than would be possible using conventional characters.

and the changes are immediately transmitted into the modified array, and hence to the screen. Figure A1.1(a)–(b) show a comparison between main and modified arrays, and Figure A1.1(c) is a sketch of the computer configuration generally employed in video work of this sort.

The computer video processor therefore constantly samples from the display file, and updates the information on the screen. It is possible to use either characters from the standard computer character set, or else write special subroutines to generate shapes drawn by vectors. Details of how to do the latter are probably best obtained from the appropriate computer video instructions. Figure A1.2 shows a step in a video representation of a cell clone growth simulation using two special shapes to indicate two types of cells.

USE OF THE LIGHT PEN

It is possible to communicate with the program by means of the light pen. We have seen in Chapter 7 that the use of arrays gives us an easy method of keeping track of the coordinates of simulated cells: by pointing at particular array positions displayed on the screen with the light pen, the computer can recognize the pen's position—groups of cells or individual cells can thus be removed, put on to the screen, or moved in position.

The video console is usually supplemented by a row of push buttons, and these can be used during the running of a program. If a button is pushed, it activates a flag in the display processor. This flag is tested for during the running of the program by using a conditional instruction of the sort 'if PB3 (pushbutton) is pushed, go to n', where n is a statement label. Push buttons may be used in conjunction with the light pen—they can be employed to 'freeze' the video ready for the light pen to be used—or to directly interact with the program in some other way. A typical interaction step might be as follows. The operator decides to remove a group of cells from a growing cell mass. The picture is 'frozen' by pushing an appropriate push button. All cells to be removed are pointed at with the light pen, and then these 'tagged' cells are removed by pushing a second push button. A third button then restarts the growth process.

PROGRAMMING THE LIGHT PEN

The X and Y coordinates of the characters representing 'cells' on the screen are not important to the programmer when he is using the computer's standard character set. Each character will be drawn with the beam ending up in the correct position to draw the next character; it is only necessary to get to the beginning of each new line at the correct point. However, when the light pen is in operation, these coordinates become very important. A picture is generated on the screen by the passage of a stream of electrons on to the phosphorescent surface of the screen, and the beam traverses each row from left to right starting at the top left-hand corner of the screen, completing one picture sweep in about 1/50th of a second. At any instant in time, the beam will be at a particular set of X, Y coordinates on the screen. These coordinates are also known to the computer, and are stored as constantly changing variables that we will designate as **LPX** (last position X) and **LPY** for X and Y respectively.

If the light pen is held at a certain point of the screen, an interrupt will be generated when the beam sweep is registered on the light pen. The programmer can then obtain **LPX** and **LPY** for the position at which the interrupt was generated, and mould those into the program as he wishes.

Now the screen is made up of a large number of raster points, perhaps as many as 1024^2—and the array element pointed at will take up more than one of these. Some method of referencing a number of raster locations to a particular array element is required. This is handled by assigning a control X, Y location to each character—the control location can then be changed back into the appropriate array position in the main array by using a specially written function.

The programmer will have prepared the program to receive the control X and Y coordinates, and a variety of steps may be carried out before transferring back to light pen operation. We have already looked at some of the operations that might be used.

Appendix 2

Further reading

Although I have reviewed and commented on various sources at the end of each chapter, it may be helpful to discuss here some of the most essential references which should be consulted by someone wanting to find out more about models in development, and particularly the systems approach to studying embryological processes.

ANALYTICAL TREATMENTS OF DEVELOPMENT

There are several categories of study here. The older school of embryologists, including Waddington, Weiss, Needham, and others, has produced much useful work which might be used as a starting point. Waddington's *Principles of Embryology* (1953) and Weiss's *Dynamics of Development: Experiments and Inferences* (1939) are worth looking at. *Modern Theories of Development* (von Bertalnffy, 1933) gives a clear overview of what problems a 'theoretical biologist' of the 1930s thought important in developmental biology. Coming more up to date, Waddington's *New Patterns in Genetics and Development* (1962) is essential reading, although someone new to embryology might try his simpler *Principles of Development and Differentiation* (1966).

In recent years several useful textbooks on general developmental biology have appeared. Grant's *Biology of Developing Systems* (1978) is a lengthy book with the accent on completeness. Ede's *An Introduction to Development Biology* (1978) is shorter, at a higher level, and is perhaps more suitable for the reader of the present text. Davenport's *Outline of Animal Development* (1979) has some discussion on the use of models in describing developing systems, although this is largely restricted to the use of gradients in specifying biological patterns.

'THEORETICAL EMBRYOLOGY'

The four volumes of *Towards a Theoretical Biology* (edited by C. H. Waddington) have many useful essays, for example: in volume 1 (1968) articles by Michie and Longuet-Higgins ('a party game model of biological replication'), Michie and Chambers ('"Boxes" as a model of pattern formation'); volume 2 (1969) Arbib ('Self producing automata'); volume 3 (1970) Kaufmann ('randomly constructed genetic nets'). Two journals of interest are the *Journal of Theoretical Biology*, and *General Systems*, both of which regularly include papers on computer modelling, often in embryology itself. At present, a lot of computer models of the type mentioned in this book never get into print at all, and others appear only in very obscure journals.

MATHEMATICS AND BIOLOGY

References to books and essays on catastrophe theory may be found in Chapter 5. Models based on more conventional mathematical ideas have been around for many years, and are discussed in Maynard-Smith's *Mathematical Ideas in Biology* (1968). An early attempt at a mathematical description of biological shapes and patterns can be found in D'Arcy Thompson's classic book *On Growth and Form*, written at the turn of the century. Although this book makes no attempt to discuss development, it represented the first real use of mathematics in studying biological shapes. Nicolis and Prigogine's book *Self Organization in Non-equilibrium Systems* reviews mathematical models of pattern formation in terms of their own mathematical ideas, but much of the mathematics will be inaccessible to the average biologist.

COMPUTERS

The first requirement here is access to the computer manuals for whatever computer that is available. It is not a good idea to tackle a model without having a basic knowledge of what the computer (and the computer language) can do. Knuth's books *The Art of Computer Programming*, especially volumes 1 and 2, are essential browsing material for self-taught programmers who don't know an accumulator from a stack.

A good general book on what computers do and how they work is *Computers made Simple* (Jacobowitz, 1966). For those who haven't learnt to program yet, the McCracken language manuals on FORTRAN and ALGOL are both very good (McCracken, 1965). Other guides to high-level languages are Woodward and Bond's *ALGOL 68-R Users Guide*, and Calderbank's *Course on Programming in Fortran IV*.

SIMULATION

Very useful here are the rather contradictory sounding *Computer Simulation Models* by J. Smith, which is a short, simple book on general methods, and Tocher's book *The Art of Simulation*. Simulation in biology is a sparser topic: for the differential equation approach, Heinmets's *Quantitative Cellular Biology* (1969) could be consulted, but there does not seem to be an equivalent to the present book, or any parallel text which takes sections of it much further. The actual models themselves can tell a lot about technique. The papers on cell aggregation simulation discussed in Chapter 8 could all be read with good effect (many are included in the volume by Mostow), and another useful exercise might be to compare the Mitolo (1973) and Ede and Law (1969) papers on limb bud simulation.

REFERENCES

von Bertalanffy, L. (1933) *Modern Theories of Development*, Harper, N.Y.
Calderbank, V. J. (1969) *A Course on Programming in Fortran IV*, Chapman and Hall, London.
Davenport, R. (1979) *Outline of Animal Development*, Addison-Wesley, N.Y.
Grant, P. (1978) *Biology of Developing Systems*, Holt Reinhart Winston, N.Y.
Ede, D. A. (1978) *An Introduction to Developmental Biology*, Blackie, Glasgow.
Ede, D. A., and Law, J. T. (1969) Computer simulation of vertebrate limb morphogenesis, *Nature*, **221**, 244–248.
Heinmets, F. (1969) *Quantitative Cellular Biology*, Dekker, N.Y.
Jacobowitz, H. (1966) *Computers made Simple*, W. H. Allen, London.
Knuth, D. (1975) *The Art of Computer Programming* (2nd edn), Addison-Wesley, Reading, Mass.
McCracken, D. D. (1965) *A Guide to Fortran IV Programming*, and *A Guide to Algol Programming*, Wiley, N.Y.
Maynard-Smith, J. (1968) *Mathematical Ideas in Biology*, Cambridge University Press, Cambridge.
Mitolo, V. (1973) A model approach to some problems of limb morphogenesis, *Acta Embryologiae Experimentalis*, 323–340.
Mostow, G. D. (ed.) (1975) *Mathematical Models for Cell Rearrangement*, Yale University Press, New Haven.
Nicolis, G., and Prigogine, I. (1977) *Self Organization in Non-equilibrium Systems*, Wiley–Interscience, N.Y.
Smith, J. (1970) *Computer Simulation Models*, Griffin, London.
Tocher, K. D. (1963) *The Art of Simulation*, English Universities Press, London.
Thompson, D. A. (1942) *On Growth and Form* (2nd edn), Cambridge University Press, Cambridge.
Turing, A. (1952) A theory of morphogenesis, *Philosophical Transactions of the Royal Society*, B, **237**, 37–72.
Waddington, C. H. (1953) *Principles of Embryology*, Allen and Unwin, London.
Waddington, C. H. (1962) *New Patterns in Genetics and Development*, Columbia University Press, N.Y.
Waddington, C. H. (1966) *Principles of Development and Differentiation*, Macmillan, London.

Waddington, C. H. (1968–70) *Towards a Theoretical Biology*, vol. 1, *Prolegomena*; vol. 2, *Sketches*; vol. 3, *Drafts*, vol. 4, *Essays*, Edinburgh University Press.
Weiss, P. (1939) *Dynamics of Development; Experiments and Inferences*.
Woodward, P. M., and Bond, S. G. (1974) *ALGOL 68-R Users Guide*, HMSO, London.

Appendix 3

Glossary of computer modelling terms

Algorithm A series of instructions or steps used in solving a specific problem.

Array An arrangement of data items each referenced by a common key. Arrays may be one- or multi-dimensional.

Array boundary An array 'edge'. If an array consists, for example, of ten items arranged sequentially, then the array boundaries are A(1) and A(10), because A(0) and A(11) are outside the array bounds.

Array coordinate The reference by which an array item is addressed, for example A(5) means item 5 of the array A.

Array element The item referenced by a specific array coordinate. Also called an array location.

Array location *See* array element.

Assignment statement The computer equivalent of the mathematical '=', except that the variable on the left-hand side is *made equal to* the term on the right-hand side. A(3) = 25 therefore has the effect of making the array element A(3) contain the number 25.

Binary notation The mathematical system used in the computer. The binary representation of numbers uses only the symbols 0 and 1. In the same way that normal decimal numbers 'carried' to the left mean that the digit concerned is multiplied by 10, the binary 'carry' means multiplication by 2. Hence:

Decimal	*Binary*
0	000
1	001
2	010

3	011
4	100
5	101

Bit pattern   A sequence of bits making up a computer word. By masking off certain bits in the word, several smaller words may be independently referenced by the computer.

Complex cells   Term used for the technique of modelling cells by using several array elements to act as parts of an individual cell.

Conceptual model   A model devised to provide a framework in which smaller-scale models, hypotheses, and facts can be fitted. Examples are the theory of relativity or of evolution.

Constraint   A device for preventing growth in a particular direction.

Control list   A one-dimensional array used of deciding which items on the main array are to be processed, and in which order.

Deterministic processes   Processes which occur in a predicted manner according to pre-existing rules.

Heuristic model   A 'trial-and-error' model, where the rules of a system are worked out by successive modifications of a model.

Input parameter   A parameter used in a particular simulation which can normally be varied in some way. An example would be the size of array used, or the number of cells put into an array prior to running a simulation.

Lattice   Another term for a two- or three-dimensional array.

List structure, processing   See not on page 108.

Look-up table   A table which allows a model to decide on its next action. Typically, the relative values of two or more variables are consulted, with the resulting actions 'built into' the computer program.

Model   A representation of a real situation in more or less abstract form.

Monte Carlo (random walk) technique	A way of using deterministic processes to statistically simulate random processes.
Neighbourhood space	Refers to the number of neighbours array elements have in an array. In a one-dimensional array each element has two neighbours, i.e. its neighbourhood space is two. In a hexagonal two-dimensional array, each cell has a neighbourhood space of six.
Random number generator	A digital computer facility whereby numbers are pseudo-randomly generated electronically.
Random Processes	Processes which occur in a random manner, for example, whether a coin shows 'heads' or 'tails' after tossing.
Simulation	The use of a computer program as an active realization of a model.
State	Simulated entities (for example 'cells') may be in several different states which determine their interaction with other parts of the program.
System	A group of parts working together in such a way as to produce an overall pattern.
Tessellation	Refers to the arrangement of elements in an array. For example, a two-dimensional array may be *tesselated* in square or hexagonal forms.

Appendix 4

A catastrophe machine

This device enables the reader to see how continuous forces can result in catastrophic jumps, and the reader is recommended to make one for himself. The use of catastrophe theory in developmental biology is described in Chapter 5.

Materials: Two elastic bands, two drawing pins, half a matchstick, a piece of cardboard, a wooden base.

Construction: Cut out a disc of cardboard of diameter x, where x is the unexpanded length of a rubber band. Attach the two elastic bands as shown in Figure A4.1, and attach the apparatus to the wooden base by means of the two drawing pins as shown in Figure A4.2. The disc should spin freely on the drawing pin.

Method: Hold the elastic band B at its end C and slowly move it randomly around an approximate area as shown with dotted lines. The disc will be usually found to move smoothly, but at certain points it will jump. At each jump point mark the position of C. A diamond-shaped curve (Figure A4.3) is soon demarcated by the points.

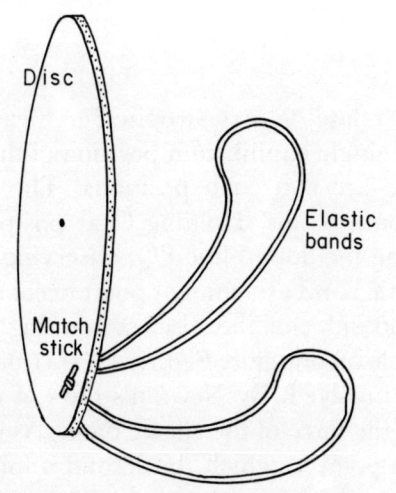

Figure A4.1   A catastrophe machine.

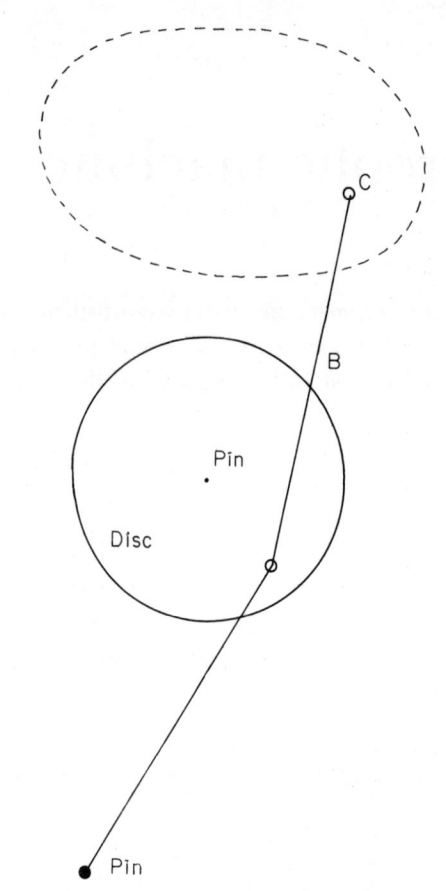

Figure A4.2 Operation of the catastrophe machine.

How does this relate to catastrophes? At each point *outside* the diamond there is a single equilibrium position of the disc, whereas *inside* the diamond there are two such positions. The following additional experiment will confirm this. Holding C at position P (Figure A4.3), slowly move it along the dotted line PQ, observing the behaviour of the disc. At P_1 the elastic band attachment point becomes stable both on the right- and left-hand sides of the disc, continuing until Q_1 is reached, when it is only stable on one side. Figure A4.4(a) shows what happens in terms of the force involved. By Newton's laws of motion, the disc will rotate to minimize the force of the elastic band. As C moves from P to Q there is therefore a point at which the second minimum becomes lower than the first, and at this point the abrupt change or 'catastrophe' will

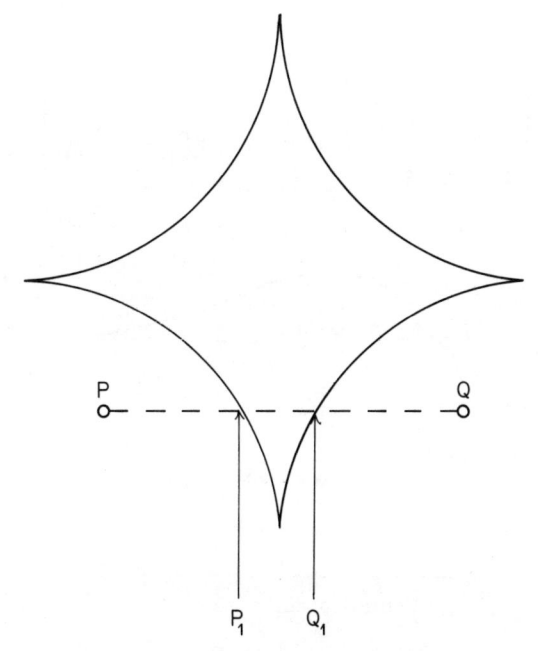

Figure A4.3 The shape outlined by the catastrophe machine.

occur. In the reverse direction, the reverse occurs, and the minima swap around.

The catastrophe machine may be analysed in terms of catastrophe theory as follows. The positional parameters u and v can be defined such that they define the state of the catastrophe machine (the derivation of these parameters may be found in Saunders, 1980—see references to Chapter 5). The location of C may be represented by a point in the u–v plane called the *control space* (Figure A4.4(b)). As u and v are altered by the movement of C, the control point traces out a path called the *control trajectory*. A second surface may be defined above the control space called the *equilibrium surface*. Unlike the control space, this surface is folded and represents the phase of the point C in topological terms. If we define a third term x as proportional to the angular change gone through by the catastrophe machine on moving C, then tracing the point C position on the equilibrium surface gives its phase at any position. Smooth variations in u and v almost always produce smooth variations in x, unless the control trajectory crosses the *bifurcation set*—the projection onto the u–v plane of the equilibrium surface folds. If C is on a surface which ends at this point, it must move to the other surface, bringing about a sudden

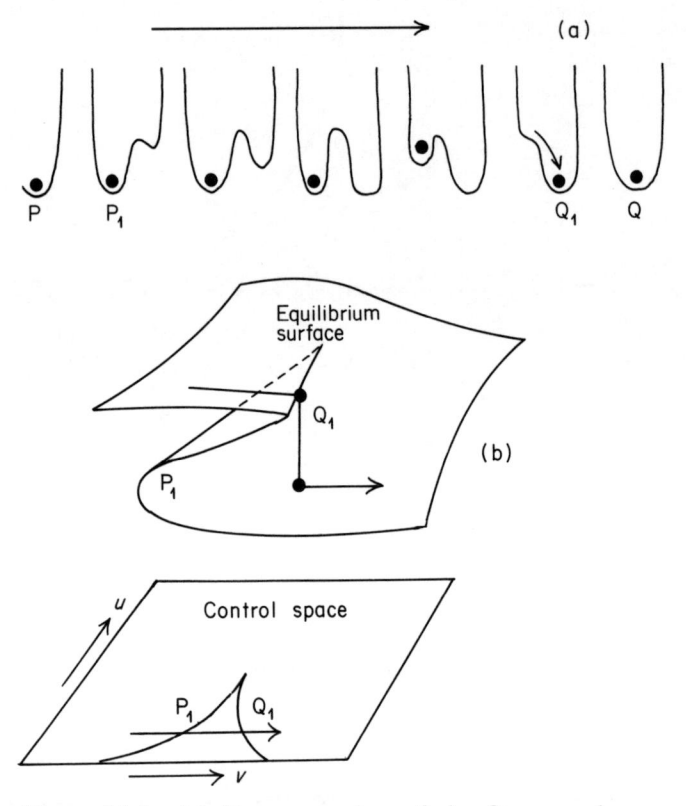

Figure A4.4 (a) Representation of the forces acting on the catastrophe machine when the pointer C is moved along the line PQ of Figure A4.3. The black dot represents the wheel position which changes at Q_1. (b) Explanation of the 'catastrophe machine' in terms of catastrophe theory.

angular change in the disc. This is a typical *cusp catastrophe* and the movement of C from P to Q can be plotted on the surfaces. As C moves along a line in the v axis, the sudden jump is predicted on the cusp at Q_1 in the reverse direction to that which occurs at P_1.

Author Index

Antonelli, P. 86, 105, 134, 135, 136, 137, 154
Apostel, L. 17
Apter, M. 17, 106, 117, 143, 152, 153
Arbib, M. 112, 153, 184, 189, 197

Baker, R. 114, 153
Becker, H. 161, 162, 163, 165, 174, 175, 182, 183
Bell, E. 54, 66
Bensam, A. 107, 153
von Bertalnffy, L. 196, 198
Bezem, J. 118, 119, 120
Bjerknes, R. 86, 88, 105, 124, 125, 128, 155
Bond, S. 197, 199
Brachet, P. 61, 67
Braverman, M. 118, 125, 153, 154, 155
Bryant, P. 161, 165, 183
Bunow, B. 58, 66
Burns, J. 9, 17

Carlsson, J. 125
Chambers, R. 197
Child, C. 16, 17
Clarke, M. 146, 156
Claxton, J. 150, 156
Codd, E. 112, 153
Cohen, M. 59, 66
Conway, J. 115, 154
Cooke, J. 65, 66
Cowe, R. 139, 141, 142, 156
Crank, J. 147, 156
Curtis, A. 133, 155

von Dalen, D. 153
Darmon, M. 61, 67
Davenport, R. 196, 198
Davis, J. 121, 155
Düchting, W. 126, 127, 154

Ede, D. 94, 105, 128, 129, 130, 132, 154, 164, 196, 198
Eden, M. 106, 114, 153
Elsdale, T. 27, 28, 29, 33, 65
Elton, R. 134, 155

Fankhauser, G. 32, 33
Fletterick, R. 103, 105, 136, 154
Frindel, E. 125, 126, 155

Gardner, M. 115, 154
Gierer, A. 58, 66, 147, 149, 156, 187
Gmitro, J. 44, 45, 46, 52, 54, 56, 66
Goel, N. 86, 105, 132, 133, 134, 136, 139, 154
Goodwin, B. 12, 13, 17, 59, 109, 110, 153
Gordon, R. 117, 118, 119, 138, 154
Grant, P. 196, 198
Grew, N. 48
Gustafson, T. 29, 30, 31, 32, 33

Hawkins, A. 121, 155
Heinmets, F. 107, 108, 153
Herman, G. 80, 114, 142, 143, 153
Holtfreter, J. 133, 155
Honda, H. 119, 120, 121, 123, 154
Hornbruch, A. 146, 156

Iverson, O. 125, 155

Jacob, F. 35, 46
Jacobowitz, H. 197, 198
Joly, G. 58, 66

Kacser, H. 9, 16, 17, 189
Kauffman, S. 57, 67, 185, 187, 189, 197
Kernevez, J. P. 58, 66
Knuth, D. 197, 198

Law, J. 94, 105, 128, 129, 130, 154, 164, 198
Lawrence, P. 41, 43, 46, 143, 146, 155, 156
Leblond, C. 125, 155
Leedale, G. 121, 155
Leith, A. 86, 105, 132, 134, 136, 139, 154
Lindenmayer, A. 80, 84, 110, 111, 112, 113, 142, 153, 154, 188
Liu, W. 142, 143, 156

Locke, M. 41
Longuet-Higgins, C. 196

McCracken, D. 197, 198
Maruyama, M. 116, 119, 154
Matela, R. 103, 105, 136, 138, 139, 154
Maynard Smith, J. 197, 198
Meinhardt, H. 58, 66, 147, 148, 149, 156, 187
Michie, D. 197
Mitchison, G. 33
Mitolo, V. 129, 131, 132, 154, 198
Monod, J. 35, 46
Mostow, G. 198

von Neumann, J. 72, 106, 111, 153, 154
Nicolis, G. 57, 65, 66, 197
Nicolson, P. 147, 156

Pearson, M. 65, 67
Postlethwait, J. 156, 161, 165, 183
Prigogine, I. 57, 65, 66, 197

Ransom, R. 7, 17, 150, 158, 174, 175, 181, 182, 183, 187
Raven, Chr. 117, 118, 119, 120, 154, 155
Rogers, T. 86, 105, 134, 135, 154
Rosen, R. 14, 17, 106, 152, 184, 189

Saunders, P. 61, 66, 205
Schneiderman, H. 156, 161, 165, 183
Scholl, C. 150, 156
Schrandt, R. 118, 124, 153, 154
Scriven, L. 44, 45, 46, 52, 54, 56, 66
Selman, G. 39
Shannon, C. 117
da Silva, L. 61, 67
Slack, J. 59, 67, 187
Smith, J. 71, 81, 197, 198
Smith, R. 33
Spemann, H. 78, 79
Stahl, W. 108, 109, 153

Steinberg, M. 132, 133, 139, 154, 155
Steward, F. 35, 46
Sugita, H. 107, 153

Thom, R. 61, 66, 188, 189
Thomas, D. 58, 66
Thompson, D. A. 12, 17, 47, 48, 49, 50, 51, 52, 65, 66, 197
Tickle, C. 134, 155
Tocher, K. 72, 73, 81, 198
Tomlinson, A. 38
Townes, P. 133, 155
Turing, A. 52, 53, 66, 143, 188

Ulam, S. 72, 77, 114, 115, 117, 119, 154

Valleron, A. J. 125, 126, 155
Vasiliev, A. V. 138, 155
Verdon, J. N. 119, 155
Vijverberg, A. 181, 183

Waddington, C. H. 14, 15, 17, 37, 139, 141, 142, 146, 156, 196, 199
Weaver, C. 117
Weiner, N. 106, 153
Weiss, P. 8, 17, 196, 199
West, J. 150, 156
Whitehead, M. 65, 67
Wilby, O. 149, 156
Wilcox, M. 24, 33
Willard, M. 86, 105, 134, 135, 154
Williams, T. 86, 88, 105, 124, 125, 128, 155
Winfree, A. 40, 46
Wiseman, L. 154
Wolpert, L. 16, 17, 29, 30, 31, 32, 117, 143, 146, 147, 153
Woodward, P. 197, 199
Wright, B. 107, 108, 153

Ycas, M. 107, 153

Zeeman, C. 61, 62, 63, 67

Subject Index

activator 58, 147, 148
algorithm 109, 110, 200
Anabaena 24, 36, 114
analogue, physical 13
animal pole 30, 32
antenna, insect 161
apical meristem 84
archenteron 32
array, border 88, 90, 200
  computer 69, 82, 142, 200
  coordinate 200
  edges 87, 88, 89
  element 164, 200
  locations 200
  two-dimensional 111, 124–126, 128, 164
assignment statement 90
autocatalysis 58, 148
automata theory 109, 110, 111, 112, 142, 188
'averaging technique' 167
axial ratio 114

bacterium 35
basal layer 125
Benard cells 44
bicaudal 75
bifurcation set 205
binary location 98
binary notation 69, 200
binding affinity 133
biologist, theoretical 196
bit packing 97, 98, 201
blastocoel 30
blastopore 30
blastula 29
butterfly 61

Calliphora 181
cancer 124, 127
cannon bones 51
carapaces, crab 51, 52
carcinogenic advantage 125
Cartesian coordinates 50
cartilage 4

catastrophe, elementary 61
  machine 203
  theory 61, 62, 64, 188, 204
cathode ray tube 191
cell, adhesion 138
  adhesiveness 31, 32
  aggregation 132
  cycle 125, 146
  division 20, 24, 25, 49, 73, 84, 90, 94, 101, 125, 126, 128, 131, 160–162, 164, 165, 178, 181, 182
  generation 90
  growth 101
  growth *in vitro* 26, 35
  interactions 8, 18, 107, 146, 154, 159
  interactions, topological model of 103
  lineage 94, 118
  mobility 128
  patterning mechanism 29
  polarity 41, 143
  population growth model 89, 191
  shape 31, 32
  sorting 136, 138
cellular automata 111
cephalin 37
chemical messengers 133
chimaeras, mammalian 150, 161
'chopping technique' 87, 89
claw 176
cleavage furrow 119
clones 96, 124, 125, 159–162, 164–167, 172, 174–178
  fragmentation 165, 168
compartment boundaries 56, 57
compiler 68
computer, analog 68, 158
  digital 68, 69, 158, 185
  language, high level 68, 73, 98
  language, simulation 71
  program 68, 70
constraint 201
control, space 205
  trajectory 205
coxa 176
creodes 15

cross-catalysis 58
cusp 61–64, 206
cuticle, hairs 41
　segmented 41
cybernetics 106, 116
cyclic AMP 61

decision procedure 108
determination 19, 71, 160
deterministic process 72, 152, 201
development, component processes 18
　genetic control of 29
　parameters 19
　theory of 187
deviation amplification 116
differentiation 19, 21, 24, 34, 35, 84, 109, 112, 121, 149, 150, 160
　eye 162
diffusion 54, 55, 76, 143, 146, 147, 150
　constant 52
　equations 72
dimensions of computer array 84, 86
directed division orientations 161
display file 191
display processor 192
dissipative structures 65
DNA 108, 109, 110, 112
double gradient theory 16
dorsal blastopore lip 78, 79
Drosophila 159, 161, 164, 176
dummy value 88

egg 18, 36
eigenfunctions 54, 55
eigenvalues 55, 56
ellipse 164, 169, 172, 173, 178
embryo, amphibian 78
　palindromic 76
embryology, theoretical 196
embryonic phaynx 162
environmental sensing 20, 24
enzyme modelling 107
epidermal cell 41, 143
epithelial layers 124
equilibrium surface 205
Erythrotrichia discigera 49, 50
Euclidean space 111
evagination movements 175, 176
evolution 185
exchange principle 134
eye, insect 161, 162

femur 176

Fibonacci series 48
fibroblast bundles 27
filament, algal 82, 84
'firing-squad synchronization problem' 142
flow chart 70
fold 61
Fourier equation 55
French flag 16, 21, 22, 23, 24, 142, 143
fruitfly 159

gases, thermodynamic behaviour of 13
gastrulation 29
　sea urchin 29
gene-action system 14, 15
genes 8, 14
genetic control 108
genetic engineering 46
genetic programme 7, 8
genotype 110
Gierer and Meinhardt model 58
gradient 20, 23–25, 33, 41, 75, 76, 143, 145–150, 196
gradients, in insect egg 76, 148
　of cell proliferation 128, 131
　of inhibitor concentration 25
　phase 59, 61
graph, plotter 190
　theory 103, 136
'ground plan' theory 79
growth, constrained 101
growth rules 89, 90

harmonic analysis 54
head, bristles 162
　chitin 162
　disc primordium 164
　discs 162, 164, 165, 175, 178
heterocyst 24, 25, 114
hit vector 90
homeostat regulator 145, 146
horn cells 119, 121, 123
Hydra 146, 149, 187
hydroid, polymorphic 124

imaginal discs 159–162, 178, 182
　leg 175, 180
　wing 56
inducer 78
information, genetic 20, 24
　theory 117
inherently precise machine 28
inhibitor 58, 147, 148
inhibitory substance 24

initiator cell   125
input   111, 112, 113
instability   56
instruction, conditional   194
invagination   29

jump instruction   70

kinematic model   111

L-systems   114
larva, insect   159, 160, 164
lattice   201
lecithin   37, 39
leg, insect   175
'levels of organization' doctrine   9, 73
life, game of   115
light pen   186, 193
limb, bud   4, 128, 131, 150
   mesenchyme cells   128
Limnaea   118, 119
line printer   190
list, computer   73
   control   90, 100, 201
list-processing techniques   108
locations, in computer   69
look-up table   73, 74, 112, 137, 201

machine code   68
mathematics, hybrid   188
matrix theory   54
membrane   101, 164
membranes, morphogenesis of   37
metamorphosis   160, 176
model   9, 201
   cell interaction   118
   computer   13, 68, 70, 71, 75, 78, 80, 81, 106, 184
   conceptual   16, 201
   descriptive   9, 14
   discrete   13
   dynamic   11, 12
   heuristic   9, 16, 80, 186, 201
   mathematical   12, 47
   paper and pencil   12, 14, 184
   positional information   16
   probabilistic   13
   rules   10
   static   12
   statistical   13
   subcellular   107
   substitute system   13, 34, 184
   topological   64, 65, 104
Monte Carlo technique   72, 73, 202

morphogen   22, 23, 52, 59, 143, 146–148, 150
morphogenesis   21, 34, 119, 121
   insect   157, 160, 161, 164
morphogenetic forces   159
motion, see-sawing   27
movements, morphogenetic   170
multi-access systems   71
myelin figures   39

neighbourhood, relation   111
   space   85, 202
nets of metabolism   185
next state   111

Olivia porphyria   139, 141
Oncopeltus   41
'onion configuration'   133, 134, 139
oscillation   59, 65
output   111, 112

parallel computation   112, 186
pattern, formation   16, 21, 114
   specification   21, 34, 107, 139, 152, 155
patterns, chemical   37, 40
   oscillatory   40
   periodic   52
Pediastrum biwae   119, 121, 123
peripodial sac   162
phenotype   14, 15, 110
physical constraints   101
pigmentation patterns   139, 142
pivot unit   101
plant systems   80
Podocoryne carnea   124
positional information   59, 146
preformation   2
probability theory   114
program, default   88
proheterocyst   25
projections, plane   121
pseudopodia   21, 30, 32
pseudo-random numbers   73, 121, 142
pupation   159, 162
push buttons (video)   194
'pushing algorithms'   94, 96
pushing model   94

random number   73, 90, 202
   starters   168
   walk technique   72
randomness   71, 202
refresh display   191

regeneration 60
  by growth 23, 24
  by remodelling 23, 24
  intercalary 59
relevence dilemma 12, 34
Rhodnius 143
RNA 108, 109, 112
RNA polymerase 109

sand model 41
search focus 129
self-assembly 184
sex determination 72
sheep, skin 150
signal field 20
signals, biochemical 187
simulation 13, 69, 70, 82
sink 22
skeletal cartilage 149
soap bubbles 41
somites, vertebrate 65
source 22
source–sink mechanisms 147
spermatozoon, human 1
spiral, growth of 117
state 202
statistical mechanics 13
stem cell 162
stolen length 124
storage tube 191
structure 9
subroutine 109, 110
swallowtail 61
switches, biochemical 59
  genetic 187
system 9, 202
  parameters 10
systems, cellular 4
  in embryology 3
  molecular 4
  organ 4
Spirogyra 49

talpid mutant 128, 131, 132
tarsal segment 176
teletype 190
tesselation, irregular 86
  regular 86, 202
tibia 176
time base 111
tissue viscosities 138
torus 88
transdetermination 19
transformation rule 110
transition function 111
transplantation, imagined disc 160
triangular cells 119, 121, 123
triangulated graph 103, 104
'trigger' hypothesis 79
trivalent map 104, 136
trochanter 176
tumour 124, 125, 128
Turing machine 108, 109, 111

Ulva 84
umbilic, elliptic 61
  hyperbolic 61
  parabolic 62
universe 10

variables, computer 69
vegetal pole 30, 32
vegetative cells 24, 25
vertebrate limb 128, 149, 150, 151
video display 141, 158, 170, 186, 190

waves, reflecting 143
wing, insect 161
wool follicles 150
word, computer 97
work of adhesion hypothesis 132, 139

zoospores 120, 121, 123, 124